AN ASSESSMENT OF THE HEALTH RISKS OF
SEVEN PESTICIDES USED FOR TERMITE CONTROL

prepared by the

COMMITTEE ON TOXICOLOGY

Board on Toxicology and Environmental Health Hazards
Commission on Life Sciences

National Academy Press
Washington, D.C.

August 1982

NOTICE: The project that is the subject of this report was approved by the Governing Board of the National Research Council, whose members are drawn from the Councils of the National Academy of Sciences, the National Academy of Engineering, and the Institute of Medicine. The members of the committee responsible for the report were chosen for their special competences and with regard for appropriate balance.

This report has been reviewed by a group other than the authors according to procedures approved by a Report Review Committee consisting of members of the National Academy of Sciences, the National Academy of Engineering, and the Institute of Medicine.

The National Research Council was established by the National Academy of Sciences in 1916 to associate the broad community of science and technology with the Academy's purposes of furthering knowledge and of advising the federal government. The Council operates in accordance with general policies determined by the Academy under the authority of its Congressional charter of 1863, which establishes the Academy as a private, nonprofit, self-governing membership corporation. The Council has become the principal operating agency of both the National Academy of Sciences and the National Academy of Engineering in the conduct of their services to the government, the public, and the scientific and engineering communities. It is administered jointly by both Academies and the Institute of Medicine. The National Academy of Engineering and the Institute of Medicine were established in 1964 and 1970, respectively, under the charter of the National Academy of Sciences.

Prepared under Contract N00014-80-C-0161 between the National Academy of Sciences and the Office of Naval Research.

Subcommittee to Assess the Hazards of Termiticides

Wendell W. Kilgore, University of California, Davis, California,
    Chairman
David W. Gaylor, National Center for Toxicological Research,
    Jefferson, Arkansas
Leonard T. Kurland, Mayo Clinic, Rochester, Minnesota
Howard I. Maibach, University of California, San Francisco, California
Edward A. Smuckler, University of California, San Francisco, California

COMMITTEE ON TOXICOLOGY

Roger O. McClellan, Lovelace Inhalation Toxicology Research Institute,
    Albuquerque, New Mexico, Chairman
Donald Ecobichon, McGill University, Montreal, Quebec, Canada
David W. Gaylor, National Center for Toxicological Research,
    Jefferson, Arkansas
Peter Greenwald, National Cancer Institute, Bethesda, Maryland
Ian T. Higgins, University of Michigan Medical Center, Ann Arbor,
    Michigan
Wendell W. Kilgore, University of California, Davis, California
Leonard T. Kurland, Mayo Clinic, Rochester, Minnesota
Howard I. Maibach, University of California, San Francisco, California
H. George Mandel, George Washington University School of Medicine,
    Washington, D.C.
Joseph V. Rodricks, Environ Corporation, Washington, D.C.
Ronald C. Shank, University of California, Irvine, California
Edward A. Smuckler, University of California, San Francisco, California
Robert Snyder, Rutgers University College of Pharmacy, Piscataway, New
    Jersey
Peter Spencer, Albert Einstein College of Medicine, Bronx, New York
Philip G. Watanabe, Dow Chemical U.S.A., Midland, Michigan

National Research Council Staff

Gary R. Keilson, Staff Scientist
Gordon W. Newell, Staff Scientist
Norman Grossblatt, Editor
Brenda Spears, Secretary

BOARD ON TOXICOLOGY AND ENVIRONMENTAL HEALTH HAZARDS

Ronald W. Estabrook, University of Texas Medical School (Southwestern), Dallas, Texas, Chairman
Philip Landrigan, National Institute for Occupational Safety and Health, Cincinnati, Ohio, Vice-Chairman
Edward Bresnick, University of Vermont, Burlington, Vermont
Theodore Cairns, DuPont Chemical Co. (retired), Greenville, Delaware
Victor Cohn, George Washington University Medical Center, Washington, D.C.
A. Myrick Freeman, Bowdoin College, Brunswick, Maine
Ronald W. Hart, National Center for Toxicological Research, Jefferson, Arkansas
Michael Lieberman, Washington University School of Medicine, St. Louis, Missouri
Richard Merrill, University of Virginia Law School, Charlottesville, Virginia
Robert A. Neal, Chemical Industry Institute of Toxicology, Research Triangle Park, North Carolina
Ian Nisbet, Clement Associates, Inc., Arlington, Virginia
John Peters, University of Southern California School of Medicine, Los Angeles, California
Liane Russell, Oak Ridge National Laboratory, Oak Ridge, Tennessee
Charles R. Schuster, Jr., University of Chicago, Chicago, Illinois

National Research Council Staff

Robert G. Tardiff, Executive Director
Gordon W. Newell, Associate Executive Director

CONTENTS

	Page
EXECUTIVE SUMMARY	1
CHLORDANE	9
HEPTACHLOR	19
ALDRIN/DIELDRIN	23
LINDANE	29
PENTACHLOROPHENOL	33
CHLORPYRIFOS	37
CONCLUSIONS AND RECOMMENDATIONS	43
Comparison of Carcinogenic Risk	43
Overall Assessment of Risks	45
Recommendations	46
TABLES	51
REFERENCES	59
BIOGRAPHIC SKETCHES OF COMMITTEE MEMBERS	73

EXECUTIVE SUMMARY

BACKGROUND

Seven pesticides are registered with the Environmental Protection Agency (EPA) for control of subterranean termites: chlordane, heptachlor, aldrin, dieldrin, lindane, pentachlorophenol, and chlorpyrifos. Chlordane is the most widely used of this group, and heptachlor the second most widely used. Before 1974, chlordane and aldrin accounted for 55 percent and 40 percent of market sales, respectively. The use of aldrin was drastically reduced after EPA cancellation hearings. However, because of recent increases in the price of chlordane, the use of aldrin to control termites is increasing, and it may soon account for 25 percent of the market. Dieldrin and lindane, although effective as termiticides, have rarely been used for this purpose. Pentachlorophenol is generally used for special applications, such as wood impregnation, and rarely for controlling subterranean termites. Chlorpyrifos is a newer product and only recently has been marketed for the control of subterranean termites.

Chlordane has been the pesticide of choice for termite control in military housing. There now are several reports of the presence of airborne chlordane in military housing long after application; such findings have caused concern over adverse health effects among residents. The problem has been related primarily to housing built on poured concrete slabs with heating and cooling ducts in or below the slabs. Contamination has occurred when the ducts cracked or when exterminators accidentally pierced the ducts during application of chlordane. As a result, the Air Force in 1978 asked the National Research Council's Committee on Toxicology, in the Board on Toxicology and Environmental Health Hazards, Commission on Life Sciences, to review the toxicity data on chlordane and to suggest an airborne concentration that could be used as a guideline in deciding whether the housing should be vacated. Two National Research Council committees had previously conducted detailed reviews on chlordane and some of the other termiticides (NRC, 1977a,b). However, neither of these studies involved an assessment of the possible health risks associated with airborne exposure of the termiticides.

The Committee on Toxicology (NRC, 1979) concluded that it "could not determine a level of exposure to chlordane below which there would be no biologic effect under conditions of prolonged exposure of families in military housing." However, it did suggest an interim airborne concentration of 5 $\mu g/m^3$, which was pragmatically determined on the basis of known concentrations of chlordane in the military housing, a review of reported health complaints of residents of contaminated housing, and a comparison with the acceptable daily intake derived from long-term animal feeding studies. The Committee also suggested that a prospective epidemiologic study of persons exposed to chlordane in military housing would help substantially in making a risk assessment.

In 1980, the Comptroller General of the United States (GAO, 1980) recommended that the EPA initiate a formal risk-benefit review of chlordane to determine whether its registered uses for subsurface termite control should be limited or canceled and whether the health of people living in housing treated with chlordane is being adversely affected. In response to these recommendations, the EPA has initiated a risk-benefit review of all seven pesticides registered for control of subterranean termites.

Faced with recurring exposures of personnel to chlordane in military housing, the Department of Defense issued an order in May 1980 prohibiting further application of chlordane for buildings with subslab or intraslab ducts. It also recommended that, where the risk and extent of possible termite damage in existing structures are considered unacceptable, studies be undertaken to determine the feasibility of sealing subslab or intraslab ducts and of renovating heating and cooling systems to use aboveground and above-slab ducts.

As a second step in the review of chlordane, the Department of Defense, through the Armed Forces Pest Management Board, requested in 1981 an independent review of the seven pesticides by the National Research Council's Committee on Toxicology. Specifically, the Committee was asked to evaluate the key information on the toxic effects of the pesticides; make a comparative assessment of the human-health risks associated with exposure to the pesticides; review the previously recommended exposure limit for airborne chlordane; and, if there are sufficient data, suggest airborne exposure limits for the other pesticides.

HEALTH EFFECTS OF THE SPECIFIC PESTICIDES

CYCLODIENES (CHLORDANE, HEPTACHLOR, ALDRIN, AND DIELDRIN)

The principal pesticides used for control of subterranean termites are the chlorinated cyclodienes--chlordane, heptachlor, aldrin, and dieldrin. These all had widespread use as pesticides until the mid-1970s, when cancellation hearings were held by the EPA. Their use since then has been severely limited, although their registration for control of termites was retained.

Acute or chronic exposure of humans to cyclodienes can produce central nervous system symptoms characterized by headache, blurred vision, dizziness, involuntary muscle movements, tremors, and seizures. Data on chronic exposure at low airborne concentrations are limited. A recent epidemiologic study of workers producing chlordane suggested that exposure has no long-term effects. However, because of shortcomings in the study and the suggestion of a trend in standard mortality ratios for deaths due to cancer in workers with increasing length of employment, more complete data are needed before firm conclusions can be reached with regard to the long-term human-health risks of chlordane and the other cyclodienes.

All four cyclodienes produced hepatocellular carcinomas in B6C3F1 mice; there was not a significant tumorigenic response in

Osborne-Mendel rats. Central nervous system effects--such as hyperexcitability, tremors, and convulsions--have also occurred in laboratory animals fed the cyclodiene termiticides. The cyclodienes are deposited in the body in fat, with biologic retention half lives on the order of days to several weeks.

Because these compounds are all persistent in the environment, they can be effective as termiticides for up to 20 yr after application.

LINDANE

Lindane is the gamma isomer of hexachlorocyclohexane. It has had widespread application as a pesticide, but its use has been severely restricted in the last several years. As a termiticide, it is primarily sprayed on the soil, and it persists in the environment for approximately 10 yr. Lindane has also been used widely in the treatment of scabies and for louse infestation.

In humans, lindane exerts its toxic action on the central nervous system. Signs of poisoning include tremors, ataxia, convulsions, and prostration. In severe cases of acute poisoning, violent tonic and clonic convulsions have occurred. Acute exposure of animals to lindane has produced diarrhea, hypothermia, hyperirritability, incoordination, and convulsions. Long-term exposure has produced nervous symptoms and fatty degeneration of the liver. Results of carcinogenicity tests in rodents have not been consistent, with both positive and negative results reported. However, it appears that the liver is one of the target organs after chronic exposure.

PENTACHLOROPHENOL

Pentachlorophenol is a wood preservative. As a termiticide, it is applied rarely to soil, but mainly directly to termite-infested wood. It is not as long-lived in the environment as the cyclodienes; its effectiveness as a termiticide after a single application lasts about 3 yr.

Symptoms of pentachlorophenol intoxication in humans include loss of appetite, respiratory difficulties, anesthesia, hyperpyrexia, sweating, dyspnea, and coma. Animals exposed to pentachlorophenol had pathologic changes in the liver and kidneys, in addition to symptoms associated with uncoupling of oxidative phosphorylation. There was no evidence of a carcinogenic effect in mice and rats given pentachlorophenol orally for 1-2 yr. Embryotoxicity and fetotoxicity have been observed in offspring of rats given purified or commercial pentachlorophenol.

CHLORPYRIFOS

Chlorpyrifos is an organophosphate pesticide with a wide variety of applications; only in recent years has it been used as a subsurface termiticide. Its effectiveness in controlling termites after a single application lasts about 4-10 yr.

Chlorpyrifos is in a different chemical class from the other termiticides discussed here, and its toxic effects also differ. The principal effect in humans and animals exposed for short periods is a reduction in plasma and red-cell cholinesterase activity. These changes have occurred after oral, dermal, and inhalation exposure.

There is no information on effects of long-term exposure of humans. Rats, mice, and dogs have been given chlorpyrifos in the diet for up to 2 yr. At the dosages tested, the only effect observed was a decrease in cholinesterase activity.

CONCLUSIONS AND RECOMMENDATIONS

To evaluate the risks associated with exposure to the seven pesticides that are available for controlling subsurface termites and to assess which of them, if any, are most appropriate for use in military housing from the standpoint of health risks, the Committee has considered several factors. These include health effects themselves and environmental end points that influence potential airborne concentrations, such as vapor pressure, persistence in the environment, and amount of material that needs to be applied for optimal effectiveness.

COMPARISON OF CARCINOGENIC RISK

Information is insufficient to determine whether carcinogenesis is the critical biologic end point in humans exposed to these pesticides, but available animal data allow some useful comparisons of carcinogenic risk. The four cyclodienes and lindane have been tested for carcinogenicity under similar experimental protocols. Each compound has produced hepatocellular carcinomas in male mice, and this end point can be used for comparing carcinogenicity.

The ED_{10} (dosage producing an incidence of liver tumors 10 percent above background) was calculated to make comparisons. ED_{10} was chosen because it is the lowest effective dosage that can be estimated with satisfactory precision, owing to the size of the experimental groups.

On the basis of the ED_{10}s, heptachlor, aldrin, and dieldrin had approximately the same carcinogenic activity and were more potent than chlordane; lindane had about one-sixth the activity of chlordane. The calculated ED_{10}s for dietary aldrin, dieldrin, heptachlor, chlordane, and lindane were 3.1, 3.6, 5.0, 16, and 103 ppm, respectively. The upper 95 percent confidence bounds on lifetime cancer risk expressed as the probability of cancer after a lifetime consumption of 1 L of water per day containing the compound at a concentration of 1 µg/L have also been estimated.

A limitation in interpreting the results of the bioassays is that the route of exposure was the diet, whereas the primary route of exposure of humans to these pesticides applied for termite control is inhalation. However, given this limitation, and on the basis of the ED_{10}s and the upper confidence bounds on lifetime cancer risk, the ranking from greatest to least risk would be: aldrin, dieldrin > heptachlor > chlordane > lindane.

The carcinogenic potential of chlorpyrifos was investigated in CD-1 mice; there did not appear to be any tumors related to administration of this pesticide. Direct comparison with the chlorinated hydrocarbons is not possible, because a different test protocol and a different strain of mice were used. However, using the highest dosage in the chlorpyrifos study, 15.8 ppm, and the same experimental conditions, one could estimate the proportion of animals that would be expected to have tumors after exposure to the other pesticides. On the basis of this analysis, chlordane and lindane would be expected to yield negative results if tested under the same conditions as chlorpyrifos. Obviously, one cannot predict from these data the carcinogenicity of chlorpyrifos at higher dosages.

Data on the carcinogenicity of pentachlorophenol are not adequate for comparisons with the other termiticides.

OVERALL ASSESSMENT OF RISKS

Because chlordane is the most widely used termiticide in military housing, the Committee used it as the reference material in making comparisons. The four cyclodiene termiticides are similar in overall health risks; each exerts toxic effects on the central nervous system. Although the data on aldrin, dieldrin, and heptachlor suggest a greater carcinogenic risk than that of chlordane, the Committee does not believe that these differences alone are sufficient to make one cyclodiene more desirable than another. The effectiveness of the cyclodienes as termiticides is fairly comparable, and they all persist in the environment for about 20 yr after application. Aldrin and dieldrin are less volatile than chlordane. Therefore, although chlordane has a smaller carcinogenic risk, the possibility of greater airborne concentrations might result in a greater hazard than would be expected from health data alone.

The carcinogenic risk of lindane is considerably less than that of chlordane on the basis of the mouse bioassay; other biologic end points, such as effects on the central nervous system, do not suggest that its toxicity differs from that of chlordane to any great degree. However, lindane is several times more volatile than chlordane and would have to be applied more often to be as effective. Therefore, there is a potential for greater airborne concentrations.

Neither pentachlorophenol nor chlorpyrifos has been shown to be carcinogenic, although they were not tested under the same conditions as the other termiticides. There are no data on humans, but pentachlorophenol has been shown to be embryotoxic and fetotoxic in rats. Chlorpyrifos differs from the other termiticides in being an organophosphate. Its toxicity is related primarily to effects on cholinesterase activity. Although the risk of chronic effects of pentachlorophenol and chlorpyrifos may not be as great as that of chlordane, there is a potential for acute effects. Because these materials need to be applied more often than chlordane to be as effective, there is a potential for higher airborne concentrations, which could increase the likelihood of acute effects.

RECOMMENDATIONS

Guidelines for Airborne Exposure

The airborne exposure limits suggested here are intended to provide guidance in estimating the health risks of the pesticides in military housing. These are not standards like those suggested by the Occupational Safety and Health Administration, and they do not guarantee absolute safety. Given the available data and the fact that under conditions of prolonged exposure of families in military housing there may be persons, such as young children, who in general are more susceptible to environmental insults, the Committee concluded that it could not determine a level of exposure to any of the termiticides below which there would be no biologic effects. The exposure limits were derived on the basis of health considerations and reflect the combined judgment of the Committee members; the feasibility of achieving the suggested airborne concentrations was not taken into account. However, every effort should be made to minimize exposure to the greatest extent feasible. In deciding which, if any, of the termiticides are most appropriate for use in military housing, one should take into account not only the toxicity and suggested airborne exposure limits, but also other factors that would influence the extent of exposure and hazard. Some of these factors are discussed briefly in this report; they include vapor pressure, persistence in the environment, and amount of material that needs to be applied. The suggested guidelines for airborne exposure should be reviewed again as soon as additional health-effects data become available.

Chlordane. The Committee on Toxicology in 1979 suggested an interim guideline for airborne chlordane in military housing of 5 $\mu g/m^3$. This was derived pragmatically on the basis of known concentrations of chlordane in military housing, a review of reported health complaints, and consideration of data from long-term animal-feeding studies. After an extensive review of the available literature on chlordane, and in the absence of definitive information on the health risks in humans and animals associated with various degrees of exposure to airborne chlordane, the Committee concludes that there are no new data that justify a change in the guideline of 5 $\mu g/m^3$ and suggests that it continue to be used. Because of the shortcomings of current data and in view of the Committee's request that more definitive data be developed, the airborne concentration of 5 $\mu g/m^3$ should be regarded as an interim guideline for exposures not exceeding 3 yr. This 3-yr period is suggested with the expectation that it will provide adequate time for the needed health data to begin developing.

Heptachlor, Aldrin, Dieldrin. The best available data for quantitatively comparing health risk are from the National Cancer Institute (NCI) mouse bioassays. On the basis of the $ED_{10}s$ estimated from these tests, heptachlor is approximately 3 times as carcinogenic and aldrin and dieldrin 5 times as carcinogenic as chlordane. Using these data, the Committee suggests interim guidelines for airborne heptachlor and airborne aldrin/dieldrin of 2 and 1 µg/m^3, respectively, for exposures not exceeding 3 yr.

<u>Lindane</u>. Carcinogenicity data on lindane are equivocal. Therefore, the Committee does not believe that this information should be used as a basis for suggesting a guideline for airborne exposure to lindane relative to exposure to chlordane. In the absence of other data for quantitative estimation of the risks of exposure to airborne lindane and because lindane is not now used to control termites in military housing, the Committee does not suggest a guideline for airborne exposure.

<u>Pentachlorophenol</u>. Because of the wide use of pentachlorophenol in ways other than as a termiticide, because it is not now used to control termites in military housing, because of its complex toxicity, and in the absence of definitive data on effects of long-term exposure to airborne pentachlorophenol, the Committee does not suggest a guideline for airborne exposure.

<u>Chlorpyrifos</u>. The Committee on Toxicology previously suggested a guideline for airborne chlorpyrifos of 100 µg/m^3, applicable for 90-d continuous exposure of Navy personnel in submarines. That guideline was based on data from ingestion. In the absence of data on effects of long-term exposure to airborne chlorpyrifos, the Committee concludes that the ingestion studies offer the best available information from which to derive a guideline. Because the population in military housing is more heterogeneous than that in submarines, the Committee suggests an <u>interim</u> guideline of 10 µg/m^3 for exposures not exceeding 3 yr.

Research Recommendations

The Committee strongly recommends that more definitive human-health data be developed for a fuller assessment of the risks of exposure to these termiticides. With increasing reports of human exposure to some of these termiticides in military and civilian housing, a clearer understanding of the potential risks becomes even more important. To provide a stronger data base on which to compare these materials more fully, the following research is recommended:

- <u>Long-Term Animal-Inhalation Studies</u>. The primary route of human exposure is inhalation, and there is a minimum of information on such exposure. Differences between routes of administration could modify the relative or absolute risks of these materials. Therefore, long-term inhalation studies of the seven termiticides are recommended. Biologic end points to investigate in these studies include neurotoxicity, carcinogenicity, effects on blood-forming tissues, and teratogenic and reproductive effects. Studies on the mechanisms of carcinogenicity (particularly for the cyclodienes) and neurotoxicity of these termiticides should also be undertaken. In addition, the role of metabolism in influencing their toxic effects needs to be examined further. If resources are insufficient for the testing of all seven materials, it is suggested that testing begin with chlordane, aldrin, and some of the noncyclodiene compounds, such as lindane and chlorpyrifos.

* **Airborne Monitoring.** An important consideration in assessing health hazards is knowledge of likely exposure concentrations. Published quantitative analytic data were available to the Committee only on chlordane; some preliminary data were available on aldrin and dieldrin. It is suggested that a program be undertaken to determine the airborne concentrations of the termiticides under conditions similar to those now used. This program should be conducted over a sufficient period to delineate the effects of such variables as time and temperature.

* **Epidemiology Data.** In several episodes in recent years, people have been exposed to chlordane in military housing. This group can be followed more readily than the general population, because their health records and whereabouts are more easily traced. It is recommended that, at a minimum, a biologic monitoring program be undertaken as a first step in a comprehensive analysis of the human health effects of chlordane. Concentrations of chlordane and its metabolites in fat, blood, and urine of persons who lived in military housing where chlordane was applied should be measured. Comparisons of these concentrations before and after exposure would provide some information on extent of exposure and on whether chlordane has accumulated in the body. The health status of these persons should also be investigated. In particular, investigators should look for signs of neurotoxicity--such as seizures, movement disorders, tremors, and chorea--and for signs of anemia and diseases of blood-forming tissues. Neurologic symptoms appear to be the most sensitive indicator of exposure in humans. It might be possible to develop a retrospective case-control study of those with suggestive neurologic symptoms and appropriate matched controls (neighbors, unaffected siblings, etc.), including correlation with concentrations of the termiticides in tissues and in indoor air. Any of the other termiticides for which there are sufficient data on exposure in military housing should be investigated in a similar fashion.

CHLORDANE

BACKGROUND INFORMATION

Chlordane is a member of a group of chemical compounds generically termed "chlorinated cyclodienes." For its use as an insecticide, and especially as a termiticide, it is available in pure and technical grades. Pure chlordane is a viscous, colorless, odorless liquid. Its solubility in water is approximately 9 µg/L at 25°C. It is highly soluble in most organic solvents, including petroleum hydrocarbons (Brooks, 1974a,b). Physical and chemical properties of chlordane are shown in Table 1.

Technical chlordane is usually sold as an emulsifiable concentrate for use in dilute water suspension as a spray, but is also supplied in powder form diluted with talc or other inert mineral dust (Aldrich and Holmes, 1969). First produced commercially in 1947, chlordane has been used primarily for termite eradication around housing foundations and for control of soil insects in corn production. Chlordane also has been an active ingredient in many household and garden pesticides (Infante et al., 1978).

The present review stems from exposure of several families living in housing on military bases; chlordane had been applied as a termiticide in and around the housing foundations several years earlier. Modifications in the heating systems and disruptions of the inground heat ducts are believed to have resulted in the entry and dispersion of chlordane through the duct systems, which led to mild to moderate systemic symptoms in residents. Such release of chlordane into the house environment can generally be controlled by repairing ducts or by closing off and circumventing the underground duct system.

There are several reports of detection of airborne chlordane in residences. Livingston and Jones (1981) sampled the air of apartments at a midwestern Air Force base for chlordane. These units had intraslab or subslab heating ducts and had been treated with chlordane by high-pressure subslab injection or by soil drench before pouring of the slab. In a preliminary survey, 13 units treated in 1978 and 43 treated before 1978 were sampled. All but five of the apartments had detectable concentrations of airborne chlordane, ranging from 0.4 to 263.5 µg/m^3. The majority had concentrations below 7.4 µg/m^3, with only four units exceeding 22 µg/m^3. In a wider survey, 435 apartments treated with chlordane were investigated. Of these apartments, 335 had measurable airborne chlordane, ranging from trace amounts to 37.8 µg/m^3. The mean chlordane concentration was approximately 2 µg/m^3. The authors found no correlation between airborne chlordane concentrations and barometric pressure, temperature, relative humidity, or year of treatment.

Lillie (1981) reported the results of an investigation of housing units at seven Air Force bases. Airborne chlordane was measured in 474 houses; 469 had the ventilation ducts in or below the slab, and five had the ducts in crawl spaces. Houses were treated with chlordane by subslab injection or exterior ditching. The concentration of airborne chlordane was below 0.6 µg/m³ in 130 of the houses, 0.7-3.4 µg/m³ in 278 of the houses, 3.5-6.5 µg/m³ in 56 of the houses, and above 6.5 µg/m³ in 10 of the houses. There was no correlation between concentration of airborne chlordane and temperature, barometric pressure, or relative humidity.

Wright and Leidy (1982) reported the results of sampling of six single-family homes with a crawl space or combination crawl space-slab construction. Three houses were treated with chlordane and three with a chlordane/heptachlor emulsion. Air samples were taken immediately after termiticide application and at 1 and 7 d and 1, 6, and 12 mo. Concentrations of chlordane ranged from 0.3 µg/m³ before treatment to 2.34-5.01 µg/m³ after treatment. The mean concentration of heptachlor was 0.01 µg/m³ before treatment and 1.0-1.8 µg/m³ after treatment.

SUMMARY OF TOXICITY INFORMATION

Several reviews on the toxicity of chlordane are available (NRC, 1977a,b; EPA, 1979a; IARC, 1979).

Acute poisoning associated with spraying, manufacture, or accidental ingestion of cyclodienes produces central nervous system symptoms, including headache, blurred vision, dizziness, slight involuntary muscular movements, tremor, sweating, insomnia, nausea, and general malaise. More severe illness is characterized by involuntary contractions of muscles or groups of muscles and by epileptiform convulsions, with loss of consciousness, urinary and fecal incontinence, disorientation, personality changes, psychic disturbances, and loss of memory. Such episodes, which may recur for 2-4 mo after cessation of exposure, are marked by abnormal encephalographic patterns (NRC, 1977b). The limited human studies with long-term exposure have not revealed any consistent or significant detrimental effect. Carcinogenicity data are essentially limited to long-term high-dose exposure of mice (rats appeared resistant).

Chlordane absorption through the skin can produce toxic effects (Gosselin et al., 1976). Dermal exposure is expected in manufacture or use of the pesticide. Absorption can range from negligible to that producing acute effects, depending on the degree of exposure. Chlordane could persist for long periods on the skin of persons using it. In one study, hexane rinsings of the hands of a former pest-control operator contained chlordane 2 yr after his last known exposure (Kazen et al., 1974).

Nisbet (1976) estimated total daily intake of chlordane on the basis of the concentration of oxychlordane stored in tissue. A chlordane intake of 9 µg/day was calculated. Nisbet also identified highly exposed segments of the general population: children, as a result of milk consumption; fishermen and their families, because of high consumption of fish and shellfish, especially freshwater fish; persons living downwind from treated fields; and persons living in houses treated with chlordane to control pests.

EFFECTS IN HUMANS

Case Histories

Several incidents involving accidental poisonings with chlordane have been reported. It is sometimes difficult to isolate the effects of chlordane from those of other factors.

Aldrich and Holmes (1969) reported a case of chlordane ingestion in a 4-yr-old girl that resulted in intermittent clonic convulsions, loss of coordination, and increased excitability. Spontaneous vomiting was not reported. The use of gastric lavage and parenteral phenobarbital was followed by disappearance of these neurologic signs and restoration of health. No residual manifestations were observed 24 h after the acute illness. Serum half-life was found to be 88 d.

Curley and Garrettson (1969) reported an accidental ingestion of an unknown quantity of chlordane by a 20-mo-old boy. Clinical features included vomiting and short interrupted seizures on admittance to the hospital. Results of liver-function tests and electroencephalographic and electrocardiographic tests were within normal limits 24 h after ingestion. Blood and fat samples were analyzed for chlordane and contained 0.27 mg/100 ml and 3.1 mg/kg of body weight, respectively. Serum alkaline phosphatase content was increased after 3 mo of observation.

Infante et al. (1978) reported a series of cases in children seen at a single pediatric hospital. Of 14 children diagnosed as having neuroblastoma between December 1974 and February 1976, five were known to have been exposed to chlordane during prenatal and postnatal development; the remaining nine had unknown chemical-exposure histories. Three cases of aplastic anemia and three cases of acute leukemia associated with chlordane exposure were diagnosed in the same period. The five patients with neuroblastoma and chlordane exposure were 32 mo to 6 yr old at the time of diagnosis. In each case, either the child or the mother (while pregnant) lived in a house that had recently been treated with chlordane for termite infestation. No information on a control series is available for comparison, and these case reports provided insufficient information to support conclusions concerning the long-term health risks of chlordane.

Occupational Exposure

Princi and Spurbeck (1951) evaluated 34 persons engaged in the manufacture of insecticides, including chlordane, and exposed through skin contact and inhalation for 11-36 mo. Physical examinations, chest x rays, urinary dilution and concentration tests, complete blood counts, routine urinalysis, hemoglobin measurement, sedimentation rate, and urinary porphyrin determinations failed to suggest any abnormalities in the men. Air was analyzed at several test locations for total chlorinated hydrocarbons. The authors concluded that no adverse effects were detected in men working in a plant with air concentrations of chlorinated hydrocarbons as high as 10 mg/m^3.

Alvarez and Hyman (1953) reported a clinical and laboratory study of 24 men 21-49 yr old who were exposed to chlordane for 2 mo to 5 yr while working in a plant where it was manufactured. Each man was given a complete examination, including blood chemistry and urine studies. One person with diabetes and two with essential hypertension were found, but their diseases were felt to be incidental findings. Seven men were found to have slight fibrotic changes in the apices of the lungs. It was believed that these lung findings were not related to the chlordane exposure. None of the 24 men had evidence of abnormalities in liver, kidneys, skin, nervous system, and blood-forming organs.

Fifteen workers exposed to airborne chlordane at 0.0012-0.0017 mg/m^3 over periods of 1-15 yr showed no evidence of toxic effects (Fishbein et al., 1964). Liver-function tests were normal in all instances, and there were no manifestations of nervous system disturbances or gastrointestinal or kidney disorders.

Repeated topical exposures to chlordane for over a year produced seizures, electroencephalographic dysrhythmia, convulsions, and twitching in a 47-yr-old nurseryman; these symptoms ceased when chlordane exposure was stopped (Barnes, 1967).

Wang and MacMahon (1979) studied a cohort of workers employed in the manufacture of chlordane and heptachlor at either of two locations between 1946 and 1976. Data obtained from death records, social security records, and employment history of 1,403 workers employed for longer than 3 mo in the production of the two compounds indicated no overall excess of deaths from cancer. While there were fewer deaths than expected overall for diseases of the circulatory system, there was a statistically significant excess of deaths from cerebrovascular disease (17 observed vs. 9.3 expected). These deaths all occurred after termination of employment and were not related to duration of exposure. Further study is needed to clarify the relationship between cerebrovascular disease and exposure to these cyclodienes.

Shindell and Associates (1980) extended the previous study and reported on all former and current employees with 3 mo or more of employment at the Marshall, Illinois, plant of Velsicol Chemical Corporation between 1946 (when the manufacture of chlordane began) and the study cutoff date of June 30, 1976. The cohort from the Marshall plant totaled 783 persons: 689 white men, 10 nonwhite men, and 84 women. In only 20 cases could it not be confirmed whether the person was still alive (classified as "status unknown"). Most of the analyses in the report are based on a comparison of the workers' health with the health of comparable segments of the U.S. population. Table 2 summarizes the causes of death in Velsicol workers with various periods of employment, compared with those in the U.S. population, and Table 3 presents the standard mortality ratios in Velsicol workers by major cause of death and period of employment. The study concluded that the mortality of these workers was not significantly higher than expected. The "healthy-worker" bias could be a problem here, inasmuch as the observed numbers of deaths are markedly lower than the expected numbers based on the U.S. population. Because of this potential bias, it would have been advantageous to compare production workers with nonproduction workers, whose exposure to chlordane was minimal.

There was a statistically significant positive trend in standard mortality ratios for cancer deaths (Table 3) in workers with increasing duration of employment. On the basis of the nonparametric randomized trend test, the statistical significance for these data is $p \leq 0.0083$ (one-sided) and $p \leq 0.0167$ (two-sided). However, the increase in the standard mortality ratios from one to 20 yr of exposure was not large. Also, these p values must be interpreted in light of the fact that a large array of data are presented (which increases the opportunity to find unusual permutations); and the actual increase in standard mortality ratios may be due to changes in the healthy-worker bias with age, duration of employment, calendar year, etc. The findings on morbidity were entirely descriptive, and the authors described them as "unremarkable." The good health index of 75 percent among plant workers was compared with that of about 66 percent found in previous studies.

Two additional features would have been desirable in this study (Shindell and Associates, 1980). First, measurements of extent of exposure presumably were not made; it may be assumed that some workers--plant operators, engineers, and maintenance workers--were more heavily exposed than white-collar workers. Second, providing data on distributions of the number of years of employment and on the year of hiring would increase knowledge of the numbers of persons "exposed" for various durations and the time since first exposure.

The authors (Shindell and Associates, 1980) concluded that there was no evidence to indicate that current or past workers at Velsicol are at an increased risk for health related problems. However, the suggestion of a trend in cancer deaths with duration of employment indicates that more complete data are needed before firm conclusions can be reached with regard to the carcinogenicity of chlordane in humans.

Public Water Contamination

Harrington et al. (1978) reported that a section of the public water system of Chattanooga, Tennessee, supplying 105 people in 42 houses, was contaminated with chlordane on March 24, 1976. Chlordane concentrations in the tap water of affected houses ranged from less than 0.1 to 92,500 ppb. In 23 houses, the concentration exceeded 100 ppb; 11 of these had concentrations greater than 1,000 ppb. In a door-to-door survey of 71 residents affected, 13 (18 percent) had symptoms compatible with mild acute chlordane toxicity, gastrointestinal symptoms (nausea, vomiting, or abdominal pain), and neurologic symptoms (dizziness, blurred vision, irritability, headache, paresthesia, or muscle dysfunction). None was hospitalized, and all recovered within 48 h after exposure with no apparent chronic damage.

EFFECTS IN ANIMALS

Acute and Short-Term Exposure

The oral LD_{50} of chordane for rats was 335 mg/kg of body weight for males and 430 mg/kg for females, and the dermal LD_{50} for rats was 840 mg/kg for males and 690 mg/kg for females (Gaines 1969). Chlordane toxicity was characterized by central nervous system stimulation. Truhaut et al. (1974) reported that the oral LD_{50}s of chlordane were 350 mg/kg for rats and 1,720 mg/kg for hamsters; 200 mg/kg was not lethal to 10 rats, nor was 1,200 mg/kg lethal to 10 hamsters. Toxicity was characterized by liver enlargement and congestion of the liver and kidneys.

There are only a few investigations of the acute inhalation toxicity of chlordane. Frings and O'Tousa (1950) exposed 60 female mice to airborne chlordane; 16 for up to 16 wk at an airborne concentration estimated to be 25-50 percent of saturation and the remainder for up to 4 d to air saturated with chlordane. The chlordane was described as being 60-75 percent chlordane and 25-40 percent unspecified related compounds. In the first group, there was loss of activity and muscle coordination within a few weeks. Liver damage was observed after 6 wk, and all mice had died by 16 wk. In the other group, most mice died within 4 d and the remainder in the next 10 d.

Ingle (1953) also investigated the effects of chlordane vapor. Twenty mice each were continuously exposed for 14 d to air saturated with chlordane. Pure and technical-grade chlordane were used, which were obtained from a source different from that reported by Frings and O'Tousa (1950). Ingle (1953) observed no deaths and no effects on the liver or central nervous system. When hexachlorocyclopentadiene was added to the chlordane, there were toxic effects on the liver and CNS and fatalities. Ingle (1953) speculated that the toxicity reported by Frings and O'Tousa (1950) may have been caused by chlordane contaminated with hexachlorocyclopentadiene.

Ambrose et al. (1953) investigated cumulative effects of chlordane at daily doses of 6.25, 12.5, 25, 50, 100, or 200 mg/kg administered to albino rats by gastric intubation for 15 d. Six groups of five rats each were given the compound in cottonseed oil. Rats given chlordane at 25 mg/kg or less in 15 daily doses showed no toxic reactions with respect to the characteristics evaluated--tremors, convulsions, and death. However, the investigators stated that the adverse effect of chlordane was noted on histologic examination of liver at all dosages. The effect consisted of the presence of abnormal intracytoplasmic bodies in the liver cells. These bodies ranged in size from small ring-like structures to large rings or skeins of eosinophilic hyaline material larger than the nuclei of the cells. The investigators further stated that there was a positive correlation between the oral dosage and the number and size of the liver-cell inclusions.

The effects of chlordane on the metabolism of estrogens in rats and mice have been studied by Welch et al. (1971). Daily intraperitoneal administration of chlordane at 2 or 5 mg/kg for 7 d reduced the uterotropic action of tritiated estrone and the concentration of tritiated estrogen in the uterus. The total-body metabolism of tritiated estrogen was increased in rats pretreated daily with intraperitoneal chlordane at 10 or 50 mg/kg for 7 d.

Bushland et al. (1948) applied chlordane as a 1.5 percent spray to cattle and hogs and as a dip to sheep and goats. The material was also sponged onto horses. The treatments were applied eight times at 4-h intervals, and observations were continued for 30 d after treatment. The principal signs of toxicity of chlordane were neurologic; the chief pathologic changes were enlarged and fatty livers and subserosal hemorrhages.

Chronic Exposure and Carcinogenicity

Lehman (1952) reported studies in which chlordane was administered to rats in the diet at 2.5, 25, or 75 ppm for 104 wk. At 75 ppm, gross effects were reported, including appetite loss, growth retardation, and unspecified signs of poisoning. Microscopic changes in the liver were reported at 25 and 2.5 ppm.

Ingle (1952) reported a study in which rats were given chlordane in the diet at 5, 10, 30, 150, or 300 ppm for 104 wk. Hyperexcitability, tremors, and convulsions were seen in rats exposed at 300 ppm by the twelfth week and at 150 ppm by the twenty-sixth week. Tremors were seen in some rats exposed at 30 ppm by about the eightieth week. Growth retardation and liver and kidney damage were reported at the two highest dosages. Only one male rat exposed at the highest dosage survived to the end of the experiment. There was a significant increase in mortality in the group exposed at 150 ppm, compared with controls, by the forty-eighth week; no differences were seen in the rats exposed at 5, 10, or 30 ppm.

The carcinogenicity data on chlordane have been extensively reviewed (Epstein, 1976; NRC 1977a,b; IARC, 1979). Chlordane was evaluated for carcinogenicity by the National Cancer Institute (1977a) and found to be carcinogenic in B6C3F1 mice, with a high incidence of hepatocellular carcinoma. In this study, chlordane was given in the diet to groups of 50 B6C3F1 mice (at 30 and 56 ppm for males and 30 and 64 ppm for females) and 50 Osborne-Mendel rats (at 204 and 407 ppm for males and 121 and 242 ppm for females) for 80 wk. There were dose-related increases in hepatocellular carcinomas in mice given chlordane--highly significant, compared with controls, at the high dosages and significant only for male mice, compared with pooled controls, at the lower dosage. In contrast with the findings in mice, hepatocellular carcinoma failed to appear at a significant incidence in rats given chlordane. It was concluded that, under the conditions of the bioassay, chlordane is carcinogenic for the liver of mice.

Epstein (1976) reported a study of technical chlordane fed to CD-1 mice for 18 mo, in which 27-86 percent of the test animals died. Animals received diets containing technical chlordane at 0, 5, 25, or 50 mg/kg of body weight; mortality at 18 mo was 27-49 percent at the lower dosages and 76 percent and 86 percent for females and males, respectively, at the higher dosage. Hepatocellular carcinoma incidences were 3/33, 5/55, 41/52, and 32/39 for males and 0/45, 0/61, 32/50, and 26/37 for females at the four dosages noted above, respectively.

Teratogenicity and Reproductive Effects

In the study of Ingle (1952) described earlier, 1 female rat from each of the test groups (chlordane in the diet at 0, 5, 10, 30, 150, or 300 ppm) was mated in the twenty-fourth or forty-eighth week. The author reported that there was no effect on litter number or size.

Mutagenicity

Pure and technical chlordane has been tested for mutagenicity in the Salmonella/microsome assay, in other microbial assays, and in the dominant-lethal test (mouse). The pure compound was negative in all these test systems, but technical chlordane was mutagenic in Salmonella without mammalian activating enzymes (Simmon et al., 1977). The mutagenicity of technical chlordane in this test system may have been due to a chemical impurity.

Ahmed et al. (1977) tested chlordane for mutagenicity in an assay based on the induction of ouabain resistance in mutant Chinese hamster V79 cells. A dose of chlordane that killed 55 percent of the cells induced mutations at a frequency of 27 per million surviving cells; in solvent controls, there was a 93 percent survival rate with 2 mutations per million survivors. Whether chlordane increased mutation frequency at concentrations that were not lethal to the test cells was not reported.

When chlordane was administered in a single dose at 50 or 100 mg/kg of body weight to Charles River CD-1 male mice that were later mated to untreated females, no dominant lethal effects were noted in the offspring (Arnold et al., 1977).

Toxicokinetics

cis-Chlordane and trans-chlordane are the primary components of the insecticide. Both are stable when held under ambient conditions or mixed with the feed of experimental animals. A single oral dose of chlordane administered to rats resulted in approximately 6 percent absorption (Barnett and Dorough, 1974). Small daily doses result in greater absorption--10-15 percent. Feeding the pure cis and trans isomers separately indicates that the cis isomer is more effectively eliminated from the rats than the trans isomer. Although the difference is not large, the data indicate that, with long-term exposure, trans-chlordane would contribute more to the body burden of the exposed animals than would cis-chlordane.

Polen et al. (1971) and Street and Blau (1972) found oxychlordane to be a primary mammalian metabolite of chlordane and to persist in adipose tissue. Street and Blau (1972) observed that the toxicity of oxychlordane was greater than that of the parent compound. Barnett and Dorough (1974) tentatively identified several hydroxylated metabolites of chlordane, including oxychlordane, in rat excreta and concluded that the metabolism of chlordane takes place via a series of oxidative enzyme reactions.

Tashiro and Matsumura (1977) attempted to isolate and identify positively the metabolic products from chlordane to establish the route of its metabolism. The major route of metabolism for both cis-chlordane and trans-chlordane is via dichlorochlordene and oxychlordane. These metabolic intermediates are further converted to two key metabolites, 1-exo-hydroxy-2-chlorochlordene and 1-exo-hydroxy-2-endo-chloro-2,3-exo-epoxychlordene, which are readily degraded further. trans-Chlordane is more readily metabolized through this route. There is yet another major metabolic route for cis-chlordane, which involves a more direct hydroxylation reaction to form 1-exo-hydroxy-dihydrochlordenes and 1,2-trans-dihydroxy-dihydro-chlordene. cis-Chlordane is more readily degraded through this latter route. As judged by a toxicity test on mosquitoe larvae, none of the metabolic end products appears to be more toxic than the original chlordane or the intermediates.

Most chlordane is excreted in the feces of rats. Only about 6 percent of the total intake is voided in the urine. Rabbits, however, provide a different pattern. Urinary elimination of chlordane in rabbits is greater than excretion in the feces. This suggests that the conjugative metabolic system is more efficient in rabbits than in rats. The patterns of excretion after inhalation of chlordane by rats follow the patterns reported for oral administration (Nye and Dorough, 1976).

Human half-life data were obtained when chlordane was accidentally ingested by a young boy (Curley and Garrettson, 1969). A whole-body half-life of 21 d was calculated--long for a drug used in therapy, but quite short for a chlorinated insecticide. Barnett and Dorough (1974) obtained a half-life of about 23 d in studies with rats fed chlordane for 56 d. The serum half-life of chlordane in a young girl was found to be 88 d by Aldrich and Holmes (1969).

EXISTING GUIDELINES AND STANDARDS

The American Conference of Governmental Industrial Hygienists (1981) has adopted a threshold limit value-time weighted average (TLV-TWA) of 0.5 mg/m^3 for chlordane in workroom air. The short-term exposure limit (15 min) was set at 2 mg/m^3. The Occupational Safety and Health Administration (1981) permissible workplace exposure limit is 0.5 mg/m^3. Both agencies noted that chlordane is absorbed through the skin and that dermal exposure should therefore be avoided. The Committee on Toxicology (NRC, 1979) suggested an interim guideline for airborne chlordane in military housing of 5 μg/m^3.

An acceptable daily dose for man has been estimated to be 0.001 mg/kg body weight (WHO/FAO, 1968). Although a limit of 3 μg/L was originally suggested for chlordane under the proposed Interim Primary Drinking Water Standards (EPA, 1975a), the final EPA regulations (EPA,

1975b) did not include a limit, in view of the cancellation proceedings under the Federal Insecticide, Fungicide, and Rodenticide Act. In 1979, EPA (1979a) estimated that exposure of chlordane in drinking water at 0.12 ng/L would result in a lifetime cancer risk of 10^{-6}. Canada has established a tentative maximal permissible limit for chlordane of 3 µg/L, applicable to raw-water supplies (EPA, 1979a).

HEPTACHLOR

BACKGROUND INFORMATION

Heptachlor, a chlorinated cyclodiene, is a white to light tan, waxy solid, which is available as a dust, dust concentrate, emulsifiable concentrate, wettable powder, and oil solution. Technical-grade heptachlor is approximately 73 percent heptachlor and 28 percent trans-chlordane and related compounds. It is a broad-spectrum insecticide that in the past was registered for use on 22 crops. After cancellation hearings held by the EPA in 1978, heptachlor's uses were restricted to corn cutworms, some seed treatments, some crops, and termite control. It is expected that by 1983 subterranean termite control will be the only use of heptachlor (Dover et al., 1981). Approximately 1-2 million pounds of heptachlor are produced annually. Whereas about 25 percent of that produced was for termite control in 1974, nearly 90 percent was for termite control in 1980.

The physical and chemical properties of heptachlor are shown in Table 1.

SUMMARY OF TOXICITY INFORMATION

Several comprehensive reviews of heptachlor are available (NRC, 1977a,b; IARC, 1974, 1979; EPA, 1979b).

EFFECTS IN HUMANS

There are very few reports on effects of human exposure to heptachlor. Symptoms of poisoning with heptachlor would be expected to be similar to those with other cyclodienes and to include headache, dizziness, incoordination, tremors, and seizures. These effects could occur from oral, dermal, or respiratory exposure. Reports of occupational exposure to chlordane (Wang and MacMahon, 1979; Shindell and Associates, 1980) included exposure to heptachlor. These are described in the section of this report on chlordane.

EFFECTS IN ANIMALS

Acute and Short-Term Exposure

The oral LD_{50} values of heptachlor have been reported to be 100 and 162 mg/kg for male and female rats, respectively (Gaines, 1969). The dermal LD_{50}s are 195 and 250 mg/kg for male and female rats, respectively. Truhaut et al. (1974) reported that the oral LD_{50}s of heptachlor were 105 mg/kg for rats and 100 mg/kg for hamsters. Heptachlor is converted in vivo by microsomal oxidation to heptachlor epoxide, a more toxic compound. The oral LD_{50} of heptachlor epoxide for rats is 62 mg/kg (Sperling and Ewinike, 1969).

Heptachlor was administered daily to sheep, pigs, and rats at 2 and 5 mg/kg of body weight for 78-86 d (Kacew et al., 1973). The authors reported hepatic necrosis in all three species, with rats the most sensitive.

Chronic Exposure and Carcinogenicity

The carcinogenicity data on heptachlor have been extensively reviewed elsewhere (Epstein, 1976; NRC, 1977a,b; IARC, 1979). Unpublished studies of Kettering Laboratories have been reported by Epstein (1976). Heptachlor of unspecified purity was given to CF rats at 1.5, 3, 5, 7, and 10 mg/kg of diet for 100 wk. Tumors were observed in both controls and test groups. Statistical analysis by Epstein (1976) showed a significant increase in multiple-site tumors in females at 5, 7, and 10 mg/kg. NCI (1977b) has reported the results of a bioassay of technical-grade heptachlor (72 percent heptachlor, 18 percent trans-chlordane, 2 percent cis-chlordane, and 2 percent nonachlor) in Osborne-Mendel rats. Groups of 50 rats of each sex were given heptachlor in the diet for 80 wk (38.9 and 77.9 ppm for males and 25.7 and 51.3 ppm for females). No hepatic tumors were observed.

Epstein (1976) also reviewed unpublished studies on the carcinogenicity of heptachlor in mice. In one study, C3H mice were given heptachlor at 10 mg/kg of diet for 2 yr. The original report concluded that heptachlor increased the incidence of benign hepatomas, but not malignant tumors, compared with controls. After histologic reevaluation (Epstein, 1976), it was concluded that there was a significant increase in hepatic carcinomas, compared with controls. In a second study reviewed by Epstein (1976), CD-1 mice were fed a mixture of 75 percent heptachlor epoxide and 25 percent heptachlor at 1, 5, and 10 mg/kg of diet for 18 mo. There was a significant increase in hepatic carcinomas in males and females at 10 mg/kg and in males at 5 mg/kg. NCI (1977b) investigated the carcinogenic potential of technical-grade heptachlor in B6C3F1 mice. Groups of 50 mice of each sex were given heptachlor in the diet for 80 wk (6.1 and 13.8 ppm for males and 9 and 18 ppm for females). There was a highly significant dose-related increase in hepatocellular carcinomas.

Teratogenicity and Reproductive Effects

Heptachlor given to rats in the diet at 6 mg/kg body weight on an unspecified schedule caused a decrease in litter size in a multigeneration study (Mestitzova, 1967). Cataracts were observed in the parents and offspring--in the latter group, soon after they opened their eyes. Results of liver- and kidney-function tests were normal in all test groups.

Mutagenicity

Heptachlor was not mutagenic when tested in several strains of Salmonella typhimurium with or without a rat-liver microsomal activation system (Marshall et al., 1976). In a dominant-lethal test, rats were fed a diet containing heptachlor at 1 or 5 mg/kg for 3 generations (Cerey et al., 1973). There were significant increases in the numbers of resorbed fetuses and increases in the numbers of abnormal mitoses in the second and third generations. A dominant-lethal test was also conducted in mice (Arnold et al., 1977). Male mice were given single doses of a mixture of 25 percent heptachlor and 75 percent heptachlor epoxide orally or intraperitoneally at 7.5 or 15 mg/kg of body weight and mated with untreated females. There was no effect on pregnancy rates, early deaths, or number of live implants per female.

Toxicokinetics

In animals, heptachlor is rapidly oxidized by microsomal enzymes to 2,3-heptachlor epoxide (Davidow and Radomski, 1953). This reaction also occurs in soils and plants (Gannon and Bigger, 1958; Gannon and Decker, 1958a). Heptachlor epoxide is more toxic than heptachlor in animals. Heptachlor is also hydrolyzed to 1-hydroxy-4,5,6,7,8,8-hexachloro-3a,4,7,7a-tetrahydro-4,5-methanoindene, which is then converted to the epoxide (Lu et al., 1975).

Heptachlor accumulates in body fat as the epoxide. Radomski and Davidow (1953) found that, when rats were given heptachlor at 30 mg/kg of diet, the maximal concentration of heptachlor epoxide in fat was reached in 2-4 wk. The concentration fell to zero 12 wk after exposure ended. Heptachlor epoxide was found in the adipose tissue in the general population of several countries (Abbott et al., 1972; Curley et al., 1973).

EXISTING GUIDELINES AND STANDARDS

The EPA (1979b) has estimated that exposure to heptachlor in drinking water at 0.023 ng/L would be associated with a carcinogenic risk of 10^{-6}.

The ACGIH (1981) has recommended a TLV-TWA of 0.5 mg/m^3 and a TLV-STEL of 2.0 mg/m^3. The OSHA (1981) permissible workplace exposure limit is 0.5 mg/m^3. Both agencies noted that heptachlor is absorbed through the skin and that dermal exposure should therefore be avoided.

ALDRIN/DIELDRIN

BACKGROUND

Aldrin and dieldrin, chlorinated cyclodienes, are broad-spectrum insecticides that are contact, stomach, and inhalation poisons. Aldrin is readily converted to dieldrin, which is considered one of the most persistent of all pesticides. These insecticides were used extensively in agriculture for over 20 yr until their use was suspended by EPA (1974); their use for termite control was retained. Although the use of aldrin and dieldrin is banned in many countries, these insecticides were being manufactured in a number of European countries at least until 1978 and are still used throughout the world.

Physical and chemical properties of aldrin and dieldrin are shown in Table 1.

SUMMARY OF TOXICITY INFORMATION

Several comprehensive reviews are available on the toxicity of aldrin and dieldrin (NRC, 1977b; IARC, 1974; EPA, 1979c; NIOSH, 1978).

EFFECTS IN HUMANS

Human poisoning from aldrin and dieldrin is characterized by major motor convulsions. Other effects include malaise, incoordination, headache, dizziness, and gastrointestinal disturbances. The seizures have developed with and without other symptoms of poisoning. Several reports have described acute intoxication, including fatalities, from ingestion of aldrin and dieldrin (Hayes, 1957, 1959; Garrettson and Curley, 1969; Taylor et al., 1979). Convulsions generally began within 30 min and lasted for up to 7 h.

Patel and Rao (1958) investigated poisoning in workers spraying dieldrin. Exposure to the pesticide was not continuous, and there were periods of varied duration of nonspraying. Of 297 workers, 20 developed adverse effects, including headaches, giddiness, muscle twitching, convulsions, and loss of consciousness. Adverse effects developed 14-154 d after exposure began, and in 17 of the workers symptoms did not develop until the second spraying period.

Abnormal EEGs have been found in persons exposed to aldrin or dieldrin, as reviewed elsewhere (Taylor et al., 1979). Avar and Czegledi-Janko (1970) examined 15 workers involved in aldrin production for 1-5 yr; of 40 workers, 15 were chosen--3 who had shown symptoms and 12 others who were chosen randomly. Physical examinations, including EEGs, took place in the last month of exposure. Nine of the workers had EEG changes from increased frequency and amplitude to changes indicative of convulsions. The three workers who had symptoms (epileptiform convulsions) were followed for 7 mo after exposure ended. By the end of this period, symptoms had subsided and EEGs were normal. No environmental monitoring was done to give an indication of the extent of exposure to aldrin.

Jager (1970) reported an extensive study of 826 chemical workers involved in production of aldrin, dieldrin, endrin, and telodrin. Workers were exposed for various periods up to 15 yr. Physical examinations took place twice a year, and blood concentrations of the pesticides were measured. The only effects reported were CNS symptoms associated with exposure to the pesticides. These symptoms generally subsided within a few weeks after removal from exposure. The author compared blood concentrations of dieldrin with symptoms of intoxication and found no clinical effects at concentrations below 0.20 µg/ml. Taking into account effects on enzyme induction in workers exposed to aldrin and dieldrin, the author reported that a blood concentration of dieldrin of 0.105 µg/ml could be considered a no-effect-level.

An earlier analysis of some of these workers was reported by Hoogendam et al. (1962, 1965). Of 225 workers exposed for less than 4 yr, 17 had convulsive episodes with slight to severe EEG changes. Symptoms were reversed after removal from exposure. The authors also reported that 17 of 70 workers exposed for 4-9 yr had some EEG changes; these changes were reversible. No impairment of liver function was found. Air samples of aldrin and dieldrin were taken during two periods; the concentrations were 0.5-8.5 mg/m^3 and 0.02-0.2 mg/m^3 for each period. Because airborne concentrations varied considerably from day to day and workers often moved from location to location in the plant, these ranges do not necessarily indicate the extent of exposure.

The effects of long-term oral exposure of dieldrin have been investigated (Hunter and Robinson, 1967; Hunter et al., 1969). Thirteen men, aged 21-52 yr, were given various dosages of HEOD (pure dieldrin); three men were given HEOD at 0 µg/d for 18 mo and 211 µg/d for 6 mo, three were given 10 µg/d for 18 mo and 211 µg/d for 6 mo, three were given 211 µg/d for 24 mo, three were given 50 µg/d for 24 mo, and one was given no HEOD. The men were also observed for 8 mo after exposure ended. No adverse effects were reported on body weight, EEGs, ECGs, hematologic tests, and serum enzymes, and there were no symptoms of intoxication. Samples of blood and adipose tissue were taken at various stages to measure the concentration of HEOD. At 18-24 mo, the blood concentration of HEOD was 5-8.6 µg/L, 1.5-12.7 µg/L, 2.9-12.8 µg/L, and 12.4-32.2 µg/L for the subjects given dosages of 50, 0 and 211, 10 and 211, and 211 µg/d, respectively. There was only a slight increase in the blood concentration, compared with controls, for the group given 10 µg/d, whereas the groups given 50 and 211 µg/d had about fourfold and tenfold increases, respectively. For the latter two groups, the HEOD blood concentration reached a plateau by about the tenth month for 50 µg/d and by the eighteenth month for 211 µg/d. The half-life of HEOD in the blood of these persons was estimated to be 369 d, with a range of 141-592 d. The concentration of HEOD in adipose tissue followed a pattern similar to that in blood; when a constant dosage was administered, it plateaued by about 9-15 mo. The ratio of the concentration of HEOD in adipose tissue to that in blood was estimated to be 136.

Aldrin and dieldrin are absorbed through the skin. Feldmann and Maibach (1974) applied ^{14}C-labeled aldrin and dieldrin to the forearms of six humans at 4 µg/cm^2 and measured urinary excretion for 6 d. For aldrin and dieldrin, respectively, 7.8 and 7.7 percent of the total dose was recovered in the urine.

EFFECTS IN ANIMALS

Acute Exposure

Aldrin and dieldrin are acutely toxic to laboratory animals by the oral, dermal, and inhalation routes. They are mildly irritating to the eye and to the skin. Both aldrin and dieldrin affect the central nervous system, producing irritability, tremors, and convulsions (Heath and Vandekar, 1964). The oral LD_{50} of aldrin for rats ranges from 46 mg/kg of body weight in peanut oil to 63 mg/kg as a wettable powder (Gaines, 1969). The dermal LD_{50} for rats ranges from 98 to 274 mg/kg. The oral LD_{50} of dieldrin for rats ranges from 38 mg/kg in peanut oil to 52 mg/kg as a wettable powder; the dermal LD_{50} ranges from 64 to 213 mg/kg.

Chronic Exposure and Carcinogenicity

Aldrin and dieldrin are carcinogenic in mice, having produced increased incidences of liver tumors (NCI, 1978a). In the NCI bioassay, 50 B6C3F1 mice of each sex were given aldrin in the diet for 80 wk (4 and 8 ppm for males and 3 and 6 ppm for females) or dieldrin in the diet at 2.5 and 5 ppm for 80 wk. All animals exhibited hyperexcitability, compared with controls. There was a significant dose-related increase in hepatocellular carcinomas in male mice exposed to either aldrin or dieldrin. Other dietary studies have also demonstrated the ability of aldrin and dieldrin to produce hepatocellular carcinomas in mice (Davis and Fitzhugh, 1962; Walker et al., 1973), and dieldrin has produced lung tumors in mice (Walker et al., 1973).

Studies of aldrin and dieldrin in rats have generally yielded negative results with regard to carcinogenicity. In NCI bioassays, 50 Osborne-Mendel rats of each sex were given aldrin in the diet at 30 and 60 ppm or dieldrin in the diet at 29 and 65 ppm for up to 80 wk (NCI, 1978a), and 24 Fischer 344 rats of each sex were given dieldrin in the diet at 2, 10, and 50 ppm for up to 105 wk (NCI, 1978b). No significant incidence of dose-related neoplasms was observed. In another study (Fitzhugh et al., 1964), Osborne-Mendel rats were given aldrin or dieldrin in the diet at 0.5, 2, 10, 50, 100, and 150 ppm for 2 yr. A high incidence of multiple-site tumors was observed only at the lower concentrations, and there was no significant increase in liver tumors.

Teratogenic and Reproductive Effects

Results of studies of the teratogenicity of aldrin and dieldrin have been equivocal. Ottolenghi et al. (1974) gave pregnant golden hamsters single oral doses of aldrin at 50 mg/kg and dieldrin at 30 mg/kg on day 7, 8, or 9 of gestation. There was an increase in fetal deaths, compared with controls, and an increase in anomalies, such as open eye, cleft palate, and webbed foot. When mice were given aldrin at 25 mg/kg or dieldrin at 15 mg/kg on day 9 of gestation, there was no effect on fetal survival or weight; however, webbed foot and cleft palate were observed.

Dix et al. (1977) administered dieldrin in corn oil (1.5 and 4 mg/kg per day) or in dimethyl sulfoxide (0.25, 0.5, and 1 mg/kg per day) to pregnant mice on days 6-14 of gestation. No teratogenic effects were observed. However, there were decreases in maternal and fetal body weights in the DMSO group and in DMSO controls. In another experiment, mice and rats were given dieldrin in peanut oil orally at 1.5, 3, and 6 mg/kg per day on days 7-16 of gestation (Chernoff et al., 1975). No teratogenic effects were observed, but at 6 mg/kg there was 41 percent mortality in rats and increased liver-to-body weight ratios and decreased weight gain in mice.

Deichmann (1972) reported that, in a six-generation mouse reproduction study, aldrin and dieldrin at 25 mg/kg per day in the diet had marked effects on fertility, gestation, viability, and lactation. At lower dosages (3, 5, and 10 mg/kg per day), the effects were not as marked.

Mutagenicity

Most of the available mutagenicity data are based on dieldrin. Dieldrin was not mutagenic in several strains of Salmonella typhimurium with or without liver activation systems (Bidwell et al., 1975; Marshall et al., 1976; Shirasu et al., 1977). With metabolic activation, aldrin was also negative in several strains of Salmonella typhimurium (Shirasu et al., 1977). Majumdar et al. (1977) reported that dieldrin was mutagenic in two of three strains of Salmonella typhimurium without activation. The mutagenic effect was more pronounced and appeared in all three strains when induced mouse-liver enzymes were added.

Toxicokinetics

Aldrin is rapidly converted to 6,7-epoxide dieldrin through epoxidation of the double bond. This is a microsomal oxidation reaction that occurs in soils, plants, and animals (Gannon and Bigger, 1958; Gannon and Decker, 1958b; Bann et al., 1956). Further reaction of dieldrin yields photodieldrin (Matsumura et al., 1970), which is more toxic than the parent compound. Hydroxy degradation products of aldrin are also found and are conjugated before being excreted.

When [^{14}C]dieldrin was administered as a single dose to rats and mice by gavage, 50-70 percent was eliminated in 1 wk (Baldwin et al.,

1972). Most of the radioactivity was detected in the feces--10 times more than in the urine.

Aldrin and dieldrin accumulate in adipose tissue. In workers exposed to aldrin and dieldrin, concentrations of dieldrin in adipose tissue and plasma were 5.67 ppm and 0.0185 ppm, respectively (Hayes and Curley, 1968). Dieldrin has also been detected in the adipose tissue of the general population in a variety of countries (Curley et al., 1973; Abbott et al., 1972; Durham, 1969).

EXISTING GUIDELINES AND STANDARDS

On the basis of carcinogenicity data and other information, EPA (1974) issued suspension notices for aldrin and dieldrin in October 1974. Evidence that was cited to support suspension included the following:

• Laboratory data indicated that two strains of mice fed a diet containing dieldrin at as low as 0.1 ppm had a significant increase in liver tumors.
• A 1973 Food and Drug Administration marketbasket survey found "measurable amounts" of dieldrin in composite samples of foods, such as dairy products.
• A 1971 EPA sampling of human fat tissues revealed detectable residues of dieldrin in 99.5 percent of them.
• On the basis of NCI methods for estimating human cancer risk, the present average dietary intake of dieldrin subjects the human population to an unacceptably high cancer risk.
• Children (particularly from birth to 1 yr of age), because their diet is high in dairy products, consume considerably more dieldrin on a body-weight basis than any other segment of the population and may therefore be at an increased risk.

The EPA (1979c) has estimated that exposure of aldrin and dieldrin in the drinking water at concentrations of 0.0046 and 0.0044 ng/L, respectively, would result in a lifetime cancer risk of 10^{-6}.

NIOSH (1978) recommended, on the basis of demonstrated potential for induction of tumors in laboratory animals, that aldrin and dieldrin be controlled and handled in the workplace as suspected carcinogens and that exposure be minimized to the greatest extent possible. NIOSH (1978) also recommended that the airborne concentration of either aldrin or dieldrin in the workplace be no higher than 0.15 mg/m^3, which was the lowest concentration detectable by validated analytic methods.

The current permissible workplace exposure limit for aldrin and dieldrin enforced by the Occupational Safety and Health Administration is 0.25 mg/m^3 (OSHA, 1981). The ACGIH (1981) has recommended for aldrin and dieldrin a TLV-TWA of 0.25 mg/m^3 and a TLV-STEL of 0.75 mg/m^3. Both agencies noted that aldrin and dieldrin are absorbed through the skin and that dermal exposure should therefore be avoided.

LINDANE

BACKGROUND INFORMATION

Lindane is the gamma isomer of hexachlorocyclohexane and is commonly referred to by the misnomer hexachlorobenzene. Technical-grade lindane contains 99 percent gamma-hexachlorocyclohexane, but also contains 1 percent of other isomers. Lindane is stable in the presence of light, heat, air, carbon dioxide, and strong acids; however, it is dehalogenated in the presence of alkali. Physical and chemical properties of lindane are shown in Table 1.

About 1 million pounds of lindane are used yearly in the United States. It is used primarily for seed treatment and to control various wood-inhabiting beetles. Very little lindane is used for termite control; for this purpose, application is primarily to soil as a surface spray, rather than by injection or trenching. Lindane is also used in the treatment of scabies and lice infestation.

Lindane was the subject of an EPA rebuttable presumption against registration proceedings. Hooker Chemical Corporation terminated production as a result of the proceedings, and there has been no reported production in the United States since 1976.

SUMMARY OF TOXICITY INFORMATION

Several comprehensive reviews of lindane are available (NRC, 1977b; IARC, 1974, 1979; EPA, 1979d).

EFFECTS IN HUMANS

Several reports have linked exposure to lindane with development of aplastic anemia, and these have been reviewed elsewhere (NRC, 1977b; IARC, 1979; Morgan et al., 1980). However, this result has not been replicated in a satisfactory animal model, and no firm causal relationship has been established between lindane exposure and anemia.

Nantel et al. (1977) investigated the health effects in 50 people who ate food contaminated with lindane. Within hours, all experienced one or more of the following: nausea, vomiting, diarrhea, headache, dizziness, and paresthesia. Twenty suffered grand mal seizures and were hospitalized for 1 d.

A series of reports (Brassow et al., 1981; Baumann et al., 1980, 1981; Tomczak et al., 1981) described the health status of 60 men (mean age, 40 yr) involved in lindane production for 1-30 yr (mean, 7.2 yr); 20 clerks and 20 dairy workers were used as controls. The authors found no increase in mortality, no impairment of the central nervous system or peripheral motor nerves, and no effects on ECGs, liver enzymes, red and white blood cells, platelet counts, or hemoglobin. Neurologic effects were measured through electromyograms, electroencephalograms, motor-nerve conduction velocity, forefinger tremors, and dexterity tracking tests. The authors noted an increase in polymorphonuclear leukocytes and prothrombin time. In 54 of the workers and in controls, blood concentrations of follicle-stimulating hormone (FSH), luteinizing hormone (LH), and testosterone were measured. There was no difference in FSH, and testosterone concentrations were slightly, but not significantly, lower than those in controls. Lindane workers had significantly higher LH concentrations. Further research is needed to determine whether these changes represent an adverse health risk.

EFFECTS IN ANIMALS

Acute Exposure

The oral LD_{50}s of lindane for mice, rats, guinea pigs, and rabbits are 86, 125-230, 100-127, and 60-200 mg/kg of body weight, respectively (Gaines, 1969). The dermal LD_{50} in rats is 1,000 mg/kg for males and 900 mg/kg for females. Signs of intoxication include diarrhea, hypothermia, hyperirritability, incoordination, and convulsions.

Chronic Exposure and Carcinogenicity

Lindane was administered daily by stomach tube to male and female rats for 6 mo at 32 mg/kg or for 17 mo at 10 mg/kg (Klimmer, 1955). Observed effects in rats given 32 mg/kg included fatty degeneration of the liver and renal tubular epithelium, vacuolization of cerebral cells, and an increase in mortality. Effects were not observed at the lower dosage. Rats were also given lindane in the diet at 2, 3, 4, 5, or 10 ppm for 12 mo, and no adverse effects were observed (Melis, 1955).

In a 2-yr feeding study (WHO/FAO, 1967), lindane was given to rats in the diet at 25, 50, or 100 ppm. At 50 and 100 ppm, there was hypertrophy of the liver, and at 100 ppm also fatty degenerative changes. No effects were observed at 25 ppm.

Several authors have investigated the carcinogenic potential of lindane. These studies have been thoroughly reviewed (NRC, 1977b; Reuber, 1979; IARC, 1979). Nagasaki et al. (1972a) fed male mice lindane in the diet at 100, 250, or 500 ppm for 24 wk; no tumors were observed at any dose. Thorpe and Walker (1973) administered lindane to CF1 mice in the diet at 400 ppm; both sexes had increases in liver tumors, compared with controls. In an NCI (1977c) bioassay, groups of 50 B6C3F1 mice of each sex were given lindane in the diet at 80 or 160 ppm for 80 wk. The incidence of hepatocellular carcinoma was significantly greater in the low-dose group, but not the high-dose group, compared with controls. The investigators concluded that lindane was not carcinogenic under the conditions of the test. Groups of 50 mice of each sex were given lindane in the diet at 12.5, 25, or 50 ppm for 80 wk (Weisse and Herbst, 1977); there was no significant increase in tumors, compared with controls.

The carcinogenicity of lindane has also been investigated in rats. Nagasaki et al. (1972b) administered lindane to groups of seven

Wistar rats in the diet at 250, 500, and 1,000 ppm. At 48 wk, one rat at the highest dose had a hepatoma and three others had hypertrophic nodules without signs of malignant tumors. In another study (NCI, 1977c), groups of 50 Osborne-Mendel rats of each sex were given lindane in the diet--at 236 or 472 ppm for males and 135 or 270 ppm for females--for 80 wk. There was no statistically significant increase in tumors in any of the test groups.

Teratogenicity and Reproductive Effects

No effects on reproductive function were observed and there was no increase in malformations in a three-generation rat study in which lindane was given in the diet at 25, 50, or 100 ppm (Palmer et al., 1978b). Teratogenic effects were not observed when lindane was given orally at 5, 10, or 15 mg/kg of body weight to rabbits on days 6-18 of gestation and to rats on days 6-16 of gestation (Palmer et al., 1978a).

When lindane was given orally for 4 mo at 0.5 mg/kg to female rats, disturbances in the estrus cycle and diminished reproductive capacity were observed (Naishtein and Leibovich, 1971). There was an increase in the incidence of stillborn pups when beagles were given lindane at 7.5 or 15 mg/kg from day 5 through the end of gestation.

Khera et al. (1979) administered Benesan (50 percent lindane) to rats orally at 0, 6.25, 12.5, or 25 mg/kg on days 6-15 of pregnancy. There were no observed effects on fetal body weight or survival and no increase in intrauterine deaths or anomalies in the groups exposed to lindane.

Mutagenicity

In a dominant-lethal assay, no mutations or reproductive effects were observed (NRC, 1977b). Lindane was not mutagenic in the host-mediated assay with Salmonella typhimurium or Serratia marcescens.

Toxicokinetics

In animals, lindane is biotransformed to chlorophenols (trichlorophenol, tetrachlorophenol, and pentachlorophenol), which are excreted free or as sulfuric acid or glucuronic acid conjugates (Engst et al., 1976; Freal and Chadwick, 1973). Chadwick et al. (1975) reported that lindane is first converted to a hexachlorocyclohexane intermediate and then to two tetrachlorophenols and three trichlorophenols.

Lindane accumulates primarily in body fat (Davidow and Frawley, 1951; Oshiba, 1972). In rats given lindane in the diet at 800 mg/kg of diet for 20 months, adipose tissue was the primary store (Davidow and Frawley, 1951). Concentrations in other tissues were lower by at least 80 percent. Lindane was eliminated from tissue within 3 wk after exposure ended. Concentrations of lindane in the general populations of many countries have ranged from 0.02 to 1.43 ppm in adipose tissue and from 0.0031 to 0.0042 ppm in blood (Durham, 1969).

EXISTING GUIDELINES AND STANDARDS

The EPA (1979d) has estimated that exposure of lindane in the drinking water at a concentration of 5.4 ng/L would result in a lifetime cancer risk of 10^{-6}.

The ACGIH (1981) has recommended a TLV-TWA of 0.5 mg/m^3 and a TLV-STEL of 1.5 mg/m^3. The OSHA (1981) permissible workplace exposure limit is 0.5 mg/m^3. Both agencies noted that lindane is absorbed through the skin and that dermal exposure should therefore be avoided.

PENTACHLOROPHENOL

BACKGROUND INFORMATION

Pentachlorophenol (commonly called penta) is a white crystal or powder formulated by chlorinating molten phenol in the presence of a catalyst. The production of this material results in other condensation products, particularly polychlorinated dioxins and furans.

Pentachlorophenol is used primarily in wood preservation. In 1977, 50 million pounds were produced, of which 80 percent was used for wood preservation, and most of the remainder was used in fungicide products applied to leather, burlap, masonry, cordage, paint, and paper. Although pentachlorophenol can be applied to soil to control termites, as a termiticide it has been limited to specific situations, such as direct application to termite-infested wood structures that cannot be easily replaced.

Physical and chemical properties of pentachlorophenol are shown in Table 1.

SUMMARY OF TOXICITY INFORMATION

Several comprehensive reviews on pentachlorophenol are available (NRC, 1977b; IARC, 1979; EPA, 1979e; Rao, 1978). Information on the toxicity of pentachlorophenol is complicated by the presence of contaminants, such as dibenzo-p-dioxins and dibenzofurans, in technical pentachlorophenol samples. Several studies have shown that much of the toxicity of pentachlorophenol can be attributed to the presence of these contaminants (Goldstein et al., 1977; Johnson et al., 1973; Kimbrough and Linder, 1975; McConnell et al., 1980; Schwetz et al., 1973). Therefore, interpretation of the results of toxicity studies with pentachlorophenol must consider the chemical composition of the pentachlorophenol samples tested.

EFFECTS IN HUMANS

The toxic effects of pentachlorophenol are associated with uncoupling of oxidative phosphorylation. Characteristic observations include loss of appetite, respiratory difficulties, hyperpyrexia, sweating, dyspnea, and coma (Menon, 1958).

Pentachlorophenol is readily absorbed through the skin. Twenty infants became ill and two died after coming into contact with nursery linens that had been washed with a laundry product containing the sodium salt of pentachlorophenol (Armstrong et al., 1969). The clinical features were characteristic of an increased metabolic rate and included tachycardia, tachypnea, fever, sweating, and acidosis.

Only a few reports in the literature have described the effects of airborne exposure to pentachlorophenol, and these did not report the extent of exposure. Eighteen workers involved in treatment of wood products with pentachlorophenol were studied by Begley et al. (1977).

Renal-clearance tests (creatinine and phosphorus) were conducted on the last day of a work period, the twentieth day of a vacation, and the fifty-first day after return from vacation. Pentachlorophenol was also measured in the blood and urine. Results of renal clearance tests were reported to be below normal in 16 of the workers before the vacation. Results returned to normal in all but six workers by the end of vacation. Clearance had again decreased 51 d after return to work, but not to the extent observed before vacation. Blood and urinary pentachlorophenol concentrations decreased by 56 percent by the end of vacation.

Bergner et al. (1965) reported five cases of poisoning in workers with respiratory and dermal exposure to pentachlorophenol. Clinical symptoms included sweating, weight loss, gastrointestinal complaints, and hyperpyrexia; there was one fatality. Urinary pentachlorophenol in the four survivors ranged from 2.4 to 17.5 mg/L.

Baader and Bauer (1951) described effects in 10 workers involved in pentachlorophenol production. Exposure to other chlorinated benzol derivatives also occurred. Symptoms included eye irritation, bronchitis, and acne. In some of the workers, acne did not develop until several months after exposure ended. In nine of the workers, acne was still present in various degrees more than a year later.

EFFECTS IN ANIMALS

Acute Exposure

The oral LD_{50} of pentachlorophenol for rats is 146-175 mg/kg, and the dermal LD_{50} is 320-330 mg/kg (Gaines, 1969). Symptoms of intoxication include accelerated respiration, vomiting, increased body temperature, tachycardia, neuromuscular weakness, and cardiac failure.

Subchronic Exposure

Several studies have reported on oral administration of pentachlorophenol for 90 d. Knudsen et al. (1974) administered pentachlorophenol in the diet to Wistar rats at 0, 25, 50, and 200 ppm (0, 1.25, 2.5, and 10 mg/kg per day, respectively). At 200 ppm, growth rate was reduced in females and liver weight was increased in males. At 50 and 200 ppm, hemoglobin was increased at 6 wk, but decreased by 11 wk. No adverse effects were observed at 25 ppm.

Kimbrough and Linder (1975) gave male rats pure or technical pentachlorophenol at 1,000 ppm. Pathologic changes in the liver were observed and were of a greater magnitude with the technical sample. Sprague-Dawley rats given technical pentachlorophenol at 3 mg/kg per day had increased liver and kidney weights (Johnson et al., 1973). When the sample was purified to reduce the amount of dioxins, no effects were seen at this dosage.

Chronic Exposure and Carcinogenicity

The potential carcinogenicity of pentachlorophenol was studied by Innes et al. (1969) and by Schwetz et al. (1978). Both studies failed to reveal any increased incidence of neoplasia. In the Schwetz et al. (1978) study, Sprague-Dawley rats were given a pentachlorophenol sample (having a lower content of nonphenolics than does technical pentachlorophenol) in the diet at 0, 1, 3, 10, and 30 mg/kg of body weight per day; males were exposed for 22 mo, and females for 24 mo. No significant increase in tumors was observed at any dosage. No toxic effects were observed in male rats at 10 mg/kg per day or less or in females at 3 mg/kg per day or less.

In the other study (Innes et al., 1969), two strains of mice were given commercial pentachlorophenol (impurities not specified) by stomach tube at 46.4 mg/kg of body weight from 7 to 28 d of age and then at 130 mg/kg in the diet for up to 78 wk. No significant increase in tumors was observed.

Teratogenicity and Reproductive Effects

Schwetz et al. (1974) analyzed the effects of purified and commercial grades of pentachlorophenol on rat embryonal and fetal development. Sprague-Dawley rats were given pure pentachlorophenol or a commercial grade at 5, 15, 30, and 50 mg/kg of body weight on days 6-15 of gestation. Embryotoxicity and fetotoxicity--as indicated by such effects as resorption, subcutaneous edema, and anomalies of the skull, ribs, and vertebrae--were observed at 15 mg/kg and greater. The no-observed-adverse-effect dosage for commercial pentachlorophenol was 5 mg/kg; delayed ossification of the skull was observed with purified pentachlorophenol at this dosage. In a second study, Schwetz et al. (1978) reported that a purified grade of pentachlorophenol given to rats in the diet at 3 mg/kg of body weight for 62 d before mating, during 15 d of mating, and throughout gestation and lactation had no effect on reproduction or neonatal survival, growth, or development. At 30 mg/kg, there was a reduction in body weight among adult rats and a decrease in neonatal survival and growth.

Larsen et al. (1975) reported a reduction in fetal weight when CD rats were given an oral dose of pentachlorophenol at 60 mg/kg of body weight on days 8-13 of gestation. Fetal deaths and resorptions were observed in hamsters when doses of 5-20 mg/kg were given on days 5-10 of gestation (Hinkle, 1973); no effects were reported at 1.25 or 2.5 mg/kg.

Mutagenicity

Vogel and Chandler (1974) reported that pentachlorophenol did not induce sex-linked recessive lethals in Drosophila melanogaster.

Toxicokinetics

Pentachlorophenol is excreted primarily in the urine in the free form (75 percent); other metabolic products include tetrachlorohydroquinone and glucuronide conjugates (Ahlborg et al., 1974; Ahlborg and Thunberg, 1978). Metabolism of pentachlorophenol is mediated by microsomal enzymes.

Pentachlorophenol is eliminated from the body fairly rapidly. Braun and Sauerhoff (1976) and Braun et al. (1977) reported that, after a dose of 10 mg/kg body weight, the plasma half-life is 15 h in rats and 78 h in monkeys. Tissue concentrations of pentachlorophenol were highest in the liver and kidneys.

EXISTING GUIDELINES AND STANDARDS

On the basis of a 90-d feeding study (Johnson et al., 1973) in which a no-observed-adverse-effect level of 3 mg/kg per day was reported, the NRC (1977b) calculated a no-adverse-effect-level in drinking water of 0.021 mg/L, for a 70-kg man consuming water at 2 L/d with the assumption that 20 percent of the total intake of pentachlorophenol would be from water. The EPA (1979e) has suggested an ambient-water quality criterion for pentachlorophenol of 0.14 mg/L.

The ACGIH (1981) has recommended a TLV-TWA of 0.5 mg/m^3 and a TLV-STEL of 1.5 mg/m^3. The OSHA (1981) permissible workplace exposure limit is 0.5 mg/m^3. Both agencies noted that pentachlorophenol is absorbed through the skin and that dermal exposure should therefore be avoided.

CHLORPYRIFOS

BACKGROUND INFORMATION

Chlorpyrifos, an organophosphate pesticide, is a white, crystalline solid marketed as Dursban and Lorsban. It has a wide variety of uses, including control of fire ants, turf and ornamental plant insects, mosquitoes, cockroaches, and termites; as a soil insecticide; seed treatment; and application to dormant and foliar plants. In August 1980, EPA granted a conditional registration for use of chlorpyrifos as a subsurface termiticide.

Physical and chemical properties of chlorpyrifos are shown in Table 1.

SUMMARY OF TOXICITY INFORMATION

EFFECTS IN HUMANS

Some studies have investigated the effects of airborne chlorpyrifos on humans. Ludwig et al. (1970) exposed volunteers to a fog containing varying concentrations of chlorpyrifos and examined the effects on plasma cholinesterase and red-cell cholinesterase activity. Decreases in plasma cholinesterase (pseudocholinesterase) activity of 84-85 percent were observed in volunteers 24 h after exposure at 132.6 mg/m^3 for 2 min and at 80.4 mg/m^3 for 4 min. Recovery to preexposure values occurred within 96 h. There was no effect on red-cell cholinesterase activity. No changes in the activity of either cholinesterase were detected 24 h after exposure at 79.1 mg/m^3 for 1 min or at 1.1 mg/m^3 for 8 min. The authors did not state whether any other changes were observed after exposure to chlorpyrifos.

Eliason et al. (1969) examined the effects on plasma cholinesterase activity in workers who had sprayed chlorpyrifos for mosquito control. One group of five men had used a 0.5 percent emulsion of chlorpyrifos for 9 d; a second group of four men had sprayed a total of 177 gal of a 0.25 percent suspension and 145 gal of a 0.5 percent emulsion for 5 d; and a third group of seven men had applied a 0.5 percent emulsion or suspension of chlorpyrifos for 2 wk.

In the first group, three of the men had decreases in plasma cholinesterase activity of 68-82 percent, compared with preexposure values; the other two had decreases of up to 52 percent and 56 percent during the 9-d spraying period, but no preexposure values were recorded. In the second group, no significant changes in plasma cholinesterase were observed during exposure to chlorpyrifos. In the third group, plasma cholinesterase activity was determined before, during, and after exposure to chlorpyrifos and compared with preexposure values and with values measured in four men not exposed to chlorpyrifos. After 1.5 wk of spraying, four of the workers had decreases in plasma cholinesterase activity of 41-91 percent, two other workers had much smaller decreases (7 percent and 15 percent), and cholinesterase activity was not measured in the seventh worker. Continued exposure for an additional 4 d did not result in further decrease in plasma cholinesterase activity. Inhibition of the enzyme was reversible: 37 d after exposure ended, the activity was similar to or greater than the preexposure activity in six of the seven workers. The four men who served as unexposed controls exhibited little change in plasma cholinesterase activity during the 2-wk observation period concomitant with the exposure period of the exposed workers. From the data supplied on the number of gallons of chlorpyrifos that each worker sprayed, it was evident that the decrease in plasma cholinesterase activity was generally related to the amount of material to which a worker was exposed. No clinical manifestations of toxicity were observed in any of the workers in the three groups exposed to chlorpyrifos.

Coulston et al. (1972) gave groups of four men chlorpyrifos orally at 0.014 mg/kg of body weight per day for 28 d, 0.03 mg/kg per day for 21 d, or 0.10 mg/kg per day for 9 d. Chlorpyrifos at 0.014 or 0.03 mg/kg per day had no significant effect on plasma and red-cell cholinesterase activity; nor were any other effects observed. A dosage of 0.10 mg/kg resulted in a decrease in plasma cholinesterase activity of 34 percent of baseline in 9 d, and treatment was stopped. Cholinesterase activity returned to preexposure values within 4 wk.

Chlorpyrifos can be absorbed through the skin to produce toxic effects; however, prolonged exposure appears necessary for appreciable absorption to take place. When chlorpyrifos in a xylene solution was applied to the skin of volunteers for 12 h at 50, 7.5, 5.0, 3.0, 1.5, and 1.0 mg/kg, no effects on plasma or red-cell cholinesterase activity were observed (Kilian et al., 1970; Pennington and Edwards, 1971). Some changes were noticed when chlorpyrifos was given for several 12-h periods with 12-h intervals between exposures. A woman given three dermal applications of chlorpyrifos at 25 mg/kg of body weight had decreases in plasma cholinesterase activity of 67 percent and 47.5 percent (compared with preexposure activity) 12 and 60 h, respectively, after the final application. Another woman was given 20 dermal applications at 5 mg/kg; she exhibited no decrease in plasma cholinesterase activity during the exposure period, but a decrease of 64.3 percent 12 h after the last exposure. No effects on red-cell cholinesterase were detected in either subject; by 7.5 d after the exposures ended, plasma cholinesterase activity had returned to normal.

EFFECTS IN ANIMALS

Acute Exposure

LD_{50} values for rats indicate that chlorpyrifos is absorbed through the skin in appreciable amounts. The dermal LD_{50} for male rats, 202 mg/kg, is in the range of the oral LD_{50}, 138-245 mg/kg (Gaines, 1969; Gray, 1965; McCollister et al; 1974). Studies in other animals have shown a wide range of sensitivity. The oral LD_{50} for chickens is 32-34.8 mg/kg (Gray, 1965; Miyazaki and Hodgson, 1972); for guinea pigs, 500 mg/kg (Gray, 1965); and for rabbits, 1,000-2,000 mg/kg (Gray, 1965). Information on the dermal effects of chlorpyrifos, although minimal, suggests that precautions should be taken to prevent skin contact. Prolonged dermal exposure of rabbits to chlorpyrifos resulted in burning, hardening of the skin, swelling, and hyperemia (WHO, 1973).

Inhalation studies in animals have examined the effects of chlorpyrifos on cholinesterase activity. Dogs, rats, sheep, and pigs exposed to a "thermal fog" or "wet-mist fog" spray containing chlorpyrifos at 145 mg/m^3 for an unspecified period showed no inhibition of cholinesterase (Gray, 1965). Rats that received chlorpyrifos at 0.007 mg/m^3 for sixteen 7-h exposures over a 21-d period exhibited no decrease in blood cholinesterase activity (Torkelson, 1965).

Chronic Exposure and Carcinogenicity

Although no data were found on the effects of long-term inhalation of chlorpyrifos, long-term feeding studies in animals have determined no-adverse-effect levels with regard to cholinesterase inhibition. Groups of rats and dogs received chlorpyrifos in the diet for 2 yr at 0-3 mg/kg of body weight per day, and plasma and red-cell cholinesterase activity was measured at various intervals (McCollister et al; 1974). Dosages of 0.1 mg/kg per day or less had no effect on plasma and red-cell cholinesterase activity in rats, whereas 0.03 mg/kg per day was the largest dose tested that had no measurable effect in dogs. Higher dietary concentrations of chlorpyrifos caused significant decreases in cholinesterase activity. No signs of toxicity were observed in any of the animals during the experiment.

In another long-term study (Warner et al., 1980), CD-1 mice were given chlorpyrifos in the diet at 0.85, 6.72, and 15.8 ppm (approximately 0.05, 0.5, and 1.5 mg/kg of body weight per day, respectively) for 105 wk. There was no significant effect on behavior, mortality, food consumption, body weight, or organ weight in treated animals, compared with controls. A variety of tumors (such as lung, liver, and lymphoreticular) and other lesions (such as inflammation and hyperplasia) were observed in control and treated mice. There appeared to be no tumors directly related to administration of chlorpyrifos.

Teratogenicity and Reproductive Effects

Pregnant CF-1 mice were given chlorpyrifos by gavage at 0, 1, 10, or 25 mg/kg on days 6-15 of gestation (Deacon et al., 1980). Severe maternal toxicity was observed at 25 mg/kg, and there was a decrease in plasma and red-cell cholinesterase activity at all dosages, compared with controls. Fetotoxicity was also reported at 25 mg/kg, and there was an increase in exencephaly at 1 mg/kg, but not at 10 or 25 mg/kg. The authors repeated the experiment with dosages of 0, 0.1, 1, and 10 mg/kg. There was a decrease in cholinesterase activity at 1 or 10 mg/kg; however, no teratogenic effects were observed at any of these dosages.

Mutagenicity

The mutagenic potential of chlorpyrifos was investigated in several strains of Salmonella typhimurium and Escherichia coli with a rat liver metabolic-activation system (Poole et al., 1977; Shirasu et al., 1976). The results were negative in each of the tests.

Toxicokinetics

Chlorpyrifos metabolism has been investigated in a few species. When rats were given a single 5-mg dose of chlorpyrifos labeled with carbon-14 by stomach tube, 88.4 percent of the total dose was recovered in the urine within 48 h (Bakke et al., 1976). Compounds identified in the urine included the glucuronide of 3,5,6-trichloro-2-pyridinol (80 percent), 3,5,6-trichloro-2-pyridinol (13 percent), and 3,5,6-trichloro-2-pyridinol glucoside (4 percent). Smith et al. (1967) found that 90 percent of the radioactivity from a 10-mg dose of chlorpyrifos labeled with chlorine-36 administered to rats by stomach tube was recoverable in 24 h--90 percent of it in the urine and 10 percent in the feces. Isolated metabolites included 3,5,6-trichloro-2-pyridyl phosphate (75-80 percent), 3,5,6-trichloro-2-pyridinol (15-20 percent), and chlorpyrifos (traces). The authors found little accumulation of chlorpyrifos in any tissue except fat. The biologic half-lives were 10 h in the liver, 12 h in the kidney, 16 h in skeletal muscle, and 62 h in fat.

The principal urinary metabolites of chlorpyrifos in cows were found to differ from those in rats. When a lactating cow was fed chlorpyrifos in the diet at 5 ppm per day for 4 d, desethylated chlorpyrifos derivatives were not found (Gutenmann et al., 1968). Instead, diethylmethyl thiophosphate (35.9 percent of total dose) and diethylmethyl phosphate (26.8 percent) were recovered in the urine, and chlorpyrifos (1.7 percent) was found in the feces. No traces of chlorpyrifos were found in the milk. In another study on lactating cows (McKellar et al., 1976), chlorpyrifos was given in the diet at 0.3, 1, 3, 10, and 30 ppm consecutively, each for 14 d. Milk and cream samples were collected throughout the experiment. Chlorpyrifos was detected in the milk at 0.01 ppm for the 30-ppm concentration and at less than 0.01 ppm for the other dietary concentrations. Higher concentrations of chlorpyrifos were found in the cream, and the residues increased with increasing dietary content. The concentrations of chlorpyrifos in the cream were less than 0.01 ppm at a dietary concentration of 3 ppm or lower, 0.03 ppm at 10 ppm, and 0.10 ppm at 30 ppm. The residues of 3,5,6-trichloro-2-pyridinol in the milk were less than 0.01 ppm at a dietary concentration of 10 ppm or less and 0.01 ppm at 30 ppm, and all cream samples had pyridinol at less than 0.025 ppm. The oxygen analogue of chlorpyrifos, which is a more potent cholinesterase inhibitor than the parent compound, was detected at less than 0.01 ppm in all milk and cream samples.

Tranformation of chlorpyrifos to 3,5,6-trichloro-2-pyridinol, one of the major metabolites in rats, results in a far less toxic product that is readily removed in the urine. The oral LD_{50}s of this hydrolysis product are 794 mg/kg for rats and over 1,000 mg/kg for chickens, compared with oral LD_{50}s of chlorpyrifos of 138-245 mg/kg for rats and 35 mg/kg for chickens (Miyazaki and Hodgson, 1972; WHO, 1973).

EXISTING GUIDELINES AND STANDARDS

ACGIH (1981) has recommended a TLV-TWA of 0.2 mg/m^3 and a TLV-STEL of 0.6 mg/m^3 for chlorpyrifos. It was noted that chlorpyrifos is absorbed through the skin and that dermal exposure should therefore be avoided. ACGIH (1980) cited data from long-term feeding studies in dogs and rats and short-term feeding studies in humans that delineated the effects on cholinesterase activity and concluded that 0.2 mg/m^3 "provides a very wide margin of safety in preventing cholinergic symptoms or organic injury."

The NRC Committee on Toxicology (1978), in a review of chlorpyrifos, used an extrapolation from ingestion to inhalation exposure to suggest an appropriate airborne concentration in submarines. If it is assumed that a given amount of chlorpyrifos has the same effect by inhalation as by ingestion and if average values for minute volume and body weight are used, the approximate no-adverse-effect level from airborne exposure can be calculated. If one assumes a minute volume of 25 L/min as the average for a man doing light to medium work for 12 h and 7 L/min while he is at rest (Altman and Dittmer, 1974) and a no-adverse-effect level for plasma and red-cell cholinesterase activity for humans of 0.03 mg/kg, tne approximate no-adverse-effect airborne concentration for a 70-kg man would be 0.1 mg/m^3.

The value of an extrapolation of this nature is obviously limited by the validity of its assumptions. However, with the limited amount of inhalation data available, it provided the best approximation of the exposure-effect relationship for chlorpyrifos; and there is a built-in safety factor, in that it assumes that 100 percent of the inhaled chlorpyrifos is absorbed.

To prevent significant decreases in cholinesterase activity, which could cause functional alterations, a 90-d continuous exposure limit for chlorpyrifos of 0.10 mg/m^3, derived from the no-adverse-effect level for ingestion, was recommended by the NRC Committee on Toxicology (1978). The recommendation assumed that exposure to chlorpyrifos would be only by inhalation.

CONCLUSIONS AND RECOMMENDATIONS

In evaluating a group of materials that have similar applications, several criteria should be considered. In the current situation, where seven pesticides are registered for control of subsurface termites, the Committee has been asked to assess which, if any, of the materials are most appropriate for use in military housing from the standpoint of health risks. Table 4 summarizes some of the important factors that the Committee has considered. Health-effects information of primary importance includes the critical responses in humans and acute and chronic effects in animals. The environmental end points in Table 4 affect the hazard of these materials by influencing the possible extent of exposure. Vapor pressure, persistence in the environment, and amount of material that needs to be applied for optimal effectiveness all influence the potential airborne concentrations in residences. Because chlordane is the most widely used termiticide in military housing, the following comparisons of risk are made with reference to it.

COMPARISON OF CARCINOGENIC RISK

There are insufficient data to determine whether carcinogenicity is the critical biologic end point in humans exposed to these pesticides, but available animal data allow some useful comparisons of risk. Chlordane, heptachlor, aldrin, dieldrin, and lindane have been tested in the National Cancer Institute carcinogenesis bioassay screening program, as discussed earlier. Similar experimental protocols were used for each. Both sexes of Osborne-Mendel rats were fed the chemicals in the diet for up to 80 wk and observed for approximately 30 wk more before sacrifice. Both sexes of B6C3F1 mice were exposed to the chemicals in the diet for up to 80 wk and then observed for another 10-13 wk. The only consistent tumorigenic response was the occurrence of hepatocellular carcinomas in the male mice. This end point was used for the comparison of the carcinogenic activity of these chemicals. The experimental data are summarized in Table 5.

The ED_{10} (dosage producing a liver tumor incidence 10 percent above background) was calculated for each termiticide with a procedure given by Crump et al. (1977). The ED_{10} was chosen because it is the lowest effective dosage that can be estimated with satisfactory precision, owing to the size of the experimental groups. Estimates of the ED_{10}s and their 95 percent confidence limits are in Table 6. Chlordane had an ED_{10} of 16 ppm. Heptachlor, aldrin, and dieldrin had approximately the same carcinogenic activity and were more potent in this bioassay than chlordane, on the basis of their ED_{10}s. Lindane showed the lowest carcinogenic activity. On the basis of the ED_{10}, chlordane is approximately 6 times (103:16) as carcinogenic as lindane.

Lifetime cancer risk and upper 95 percent confidence bounds on lifetime cancer risk at low doses have also been statistically estimated for these pesticides (NRC, 1977b). The upper bounds on lifetime cancer risk are shown in Table 7 and are expressed as probabilities of cancer after lifetime daily consumption of 1 L of water containing 1 µg of the compound. They have been corrected for species conversion on the basis of dose per unit surface area; other conversion factors, such as dose per kilogram of body weight, also are available. The estimates of cancer risk were derived by using a probabilistic multistage model. At low doses, this model is often mathematically equivalent to the linear or single-hit model. If the precise mechanism of carcinogenesis is represented by a threshold or log-normal dose-response relationship, the multistage model may considerably overestimate the risk at low doses. However, this possibility cannot be reasonably quantified. Use of different risk-assessment procedures could yield quite different estimates, and they rely to some extent on subjective judgment about the nature of the dose-response curves beyond the experimental data and on assumptions about the unknown effects of potential species differences in metabolism, physiology, and carcinogenic processes. Because of these uncertainties, some are of the opinion that the magnitude of cancer risk cannot be reliably estimated (NRC, 1977a).

An additional limitation in interpreting the carcinogenicity data is that the route of exposure in the bioassays is ingestion, whereas inhalation is the primary route of exposure of humans from pesticide use in termite control. However, within this limitation and with the ED_{10}s and upper 95 percent confidence bounds on lifetime cancer risk, one can assess the relative carcinogenic risk of these pesticides. The ranking from greatest risk to least risk would be: aldrin, dieldrin > heptachlor > chlordane > lindane.

The carcinogenic potential of chlorpyrifos was investigated in CD-1 mice to which the compound was administered in the diet for 2 yr. There were no tumors related to administration of chlorpyrifos. Direct comparison of the chlorpyrifos data with those on the chlorinated hydrocarbon termiticides tested in the NCI bioassay is not possible, because different strains of mice were used and a maximum tolerated dose was not used. One could make some comparison by estimating the hepatocellular-carcinoma rates above background in B6C3F1 mice for these compounds at the highest dosage of the chlorpyrifos study. At 15.8 ppm, which did not produce tumors in CD-1 mice exposed to chlorpyrifos, the following estimates would be obtained:

Compound	Estimated proportion of animals with tumors	95 percent confidence limits
Lindane	0.016	0-0.03
Chlordane	0.10	0.06-0.14
Heptachlor	0.6	0.4-0.8

Aldrin and dieldrin are not included, because 15.8 ppm is above the maximum tolerated dose. At 15.8 ppm, heptachlor would be expected to yield a high rate of hepatocellular carcinoma. The estimated tumor rate of 10 percent for chlordane would not be expected to be significantly different from that in controls, if there were 50 animals per exposure group. Lindane would possibly not produce more tumors than in controls at 15.8 ppm. Therefore, under the test conditions used for chlorpyrifos, chlordane and lindane might also have given negative results for carcinogenicity. Obviously, one cannot predict from these data the behavior of chlorpyrifos in B6C3F1 mice or in CD-1 mice at higher dosages.

Data on the carcinogenicity of pentachlorophenol are not adequate for comparison with the other termiticides.

OVERALL ASSESSMENT OF RISKS

The four cyclodiene termiticides are similar in overall health risk. Exposure of humans to cyclodienes produces effects on the central nervous system, characterized by seizures, tremor, and incoordination. CNS effects have been observed in cases of acute poisoning, as well as after exposure for longer periods. But data on chronic exposure are sparse, and there is no information on effects of long-term exposure at low airborne concentrations. A recent epidemiologic study of workers producing chlordane suggested that exposure has no long-term effects. However, there were shortcomings in the study, and more complete data are needed before firm conclusions with regard to long-term health risks can be reached.

Each of the four cyclodienes has produced hepatocellular carcinomas in mice. The carcinogenic risk varies to some extent; NCI mouse bioassays have suggested that aldrin, dieldrin, and heptachlor have a greater carcinogenic risk than chlordane. However, the difference is not sufficient to make one termiticide more desirable than another solely on the basis of health risk. In addition, these three pesticides are no more effective, and may even be less effective, for control of termites than chlordane. Another factor to be considered in comparing termiticides is the expected airborne concentration after application. Chlordane is the only one of these termiticides on which published analytic data were available to the Committee, although some preliminary data were available on aldrin and dieldrin. Vapor pressure can provide some additional indication of expected airborne concentration. Aldrin and dieldrin are less volatile than chlordane. Therefore, although chlordane has a smaller carcinogenic risk, the possibility of greater airborne concentration might result in a greater hazard than would be expected from health data alone.

Lindane also has toxic effects on the central nervous system in humans. Signs of poisoning include tremors, ataxia, convulsions, and prostration. Violent tonic and clonic convulsions have occurred in severe cases. In animals, exposure to lindane has produced diarrhea, hypothermia, hyperirritability, incoordination, and convulsions. Extended exposure has produced nervous symptoms and fatty degeneration of the liver. Carcinogenicity data on lindane have been equivocal. On the basis of the ED_{10} estimated from the NCI mouse bioassay, the carcinogenic risk of lindane is considerably less than that of chlordane. However, it is much more volatile than chlordane and persists in the environment only about half as long as chlordane. Therefore, lindane would have to be applied more often to be as effective, and this would increase the potential for human exposure to it.

In humans, pentachlorophenol also affects the central nervous system. Effects include loss of appetite, respiratory difficulty, anesthesia, hyperpyrexia, sweating, dyspnea, and coma. Animals exhibit symptoms associated with uncoupling of oxidative phosphorylation. In addition, pathologic changes in the liver and kidneys have been observed. Unlike the other chlorinated hydrocarbons discussed here, pentachlorophenol has not been shown to be carcinogenic (although it was not tested under the same conditions). However, it has produced embryotoxicity and fetotoxicity in rats. Pentachlorophenol is not as persistent as chlordane and therefore would have to be applied more often to be as effective. This would increase the likelihood of toxic effects, particularly from acute exposure.

Chlorpyrifos presents a somewhat different situation. Being an organophosphate, it is in a different chemical class from the other pesticides discussed here. The principal effect in humans and animals exposed for short periods is a reduction in plasma and red-cell cholinesterase activity. There is no information on effects of long-term exposure of humans. The only observed effect in animals exposed for long periods was a decrease in cholinesterase activity. Carcinogenicity of chlorpyrifos has not been observed in various animal species tested (although it was not tested under the same conditions as the other termiticides). Although the risk of chronic effects may not be as great as that for chlordane, a potential for acute effects exists. Because chlorpyrifos is not as persistent as chlordane and needs to be applied more often to be effective, there is a potential for higher airborne concentrations, which could increase the likelihood of acute effects.

RECOMMENDATIONS

GUIDELINES FOR AIRBORNE EXPOSURE

The airborne exposure limits suggested here are intended to provide guidance in estimating the health risks of the pesticides in military housing. These are not standards like those suggested by the Occupational Safety and Health Administration, and they do not guarantee absolute safety. Given the available data and the fact that under conditions of prolonged exposure of families in military housing there may be persons, such as young children, who in general are more susceptible to environmental insults, the Committee concluded that it could not determine a level of exposure to any of the termiticides below which there would be no biologic effects. The exposure limits were derived on the basis of health considerations and reflect the combined judgment of the Committee members; the feasibility of achieving the suggested airborne concentrations was not taken into account. However, every effort should be made to minimize exposure to the greatest extent feasible. In deciding which, if any, of the termiticides are most appropriate for use in military housing, one should take into account not only the toxicity and suggested airborne exposure limits, but also other factors that would influence the extent of exposure and hazard. Some of these factors are discussed briefly in this report; they include vapor pressure, persistence in the environment, and amount of material that needs to be applied. The suggested guidelines for airborne exposure should be reviewed again as soon as additional health-effects data become available.

Chlordane

In 1979, the Committee on Toxicology (NRC, 1979) suggested an _interim_ guideline for airborne chlordane of 5 µg/m^3. This was a pragmatically derived concentration that was based on known concentrations of chlordane in Air Force housing, on a review of reported health complaints of residents of contaminated homes, and on a comparison with the acceptable daily intake derived from long-term animal-feeding studies. The Committee stressed that additional data were needed to assess fully the human health risks of exposure to chlordane. In particular, the Committee recommended in 1979 an epidemiologic study of persons who had resided in military housing that had been contaminated in several previous episodes. The Committee recommended that attempts be made to correlate health effects with blood concentrations of chlordane.

From an extensive review of the literature, the Committee has concluded that there are no new data that justify a change in the currently suggested airborne concentration of chlordane. The one recent epidemiologic study (Shindell and Associates, 1980) had limitations, including a lack of reporting of airborne concentrations of chlordane and a sample size that limited the ability to detect small responses in the population. Although this study suggested a trend in standard mortality ratios for deaths due to cancer in workers with increasing duration of employment and chlordane was shown to produce hepatomas in mice (NCI, 1977a), these reports do not provide information on the health risks in humans and animals associated with various degrees of exposure to airborne chlordane. Because of the shortcomings of current data and in view of the request that more definitive data be developed, the airborne concentration for chlordane of 5 µg/m^3 should be regarded as an _interim_ guideline for exposures not exceeding 3 yr. This 3-yr period is suggested with the expectation that it will provide adequate time for the needed health data to begin developing.

Heptachlor, Aldrin, Dieldrin

On the basis of the ED_{10}s estimated from the NCI mouse bioassay data, heptachlor is approximately 3 times as carcinogenic and aldrin and dieldrin 5 times as carcinogenic as chlordane. For quantitative comparison of health risks, these appear to be the best available data. Using these data, the Committee suggests <u>interim</u> guidelines for airborne heptachlor and airborne aldrin/dieldrin of 2 and 1 µg/m^3, respectively, for exposures not exceeding 3 yr.

If it is assumed that all this inhaled material would be absorbed into the blood, the guidelines for airborne exposure compare favorably with the blood concentrations associated with no observed adverse effects. For example, with 100 percent absorption and assuming a blood volume of 5 L and a daily volume of inspired air of 20 m^3, exposure of dieldrin at 1 µg/m^3 would result in a blood concentration of 4 µg/L. This is probably a conservative estimate, in that it is not likely that 100 percent of the inhaled material would be absorbed in the blood. Even with the possibility of accumulation from repeated exposure taken into account, Hunter <u>et al</u>. (1969) reported that, in volunteers given HEOD (pure dieldrin) at 50 µg/d for up to 2 yr, the blood concentration was only 5.0-8.6 µg/L. A blood concentration of 4 µg/L is considerably less than the concentration of 105 µg/L reported to be a no-adverse-effect concentration for humans (Jager, 1970).

<u>Lindane</u>

Because the carcinogenicity data on lindane are equivocal, the Committee does not believe that this information should be used as a basis for suggesting a guideline for airborne exposure to lindane relative to exposure to chlordane. In the absence of other data for estimation of the risks of exposure to airborne lindane and because lindane is not now used to control termites in military housing, the Committee does not suggest a guideline for airborne exposure.

<u>Pentachlorophenol</u>

Pentachlorophenol is rarely used to control subsurface termites through soil application or injection, but rather is applied directly to wood. Because of the wide use of pentachlorophenol in ways other than as a termiticide, because it is not now used to control termites in military housing, because of its complex toxicity, and in the absence of definitive information on effects of long-term exposure to airborne pentachlorophenol, the Committee does not suggest a guideline for airborne exposure.

<u>Chlorpyrifos</u>

The Committee previouly suggested a guideline for airborne chlorpyrifos of 100 µg/m^3, applicable for 90-d continuous exposure of Navy personnel in submarines (NRC, 1978). That guideline was based on data from ingestion. In the absence of data on effects of long-term exposure to airborne chlorpyrifos, the Committee concludes that the ingestion studies offer the best available information from which to derive a guideline. However, because the population in military housing is more heterogeneous than that in submarines, the

Committee suggests an <u>interim</u> guideline of 10 μg/m^3 for exposures not exceeding 3 yr. The Committee recognizes the limitations of extrapolating from ingestion to inhalation exposure, such as the potential wide variability of the respiratory rate in a heterogeneous population and the absence of data with which to estimate the absorption factor for inhaled chlorpyrifos, but believes it to be the best available approach.

RESEARCH RECOMMENDATIONS

The Committee strongly recommends that more definitive human-health data be developed for a better assessment of the risks of exposure to these termiticides. With increasing reports of human exposure to some of these termiticides in military and civilian housing, a clearer understanding of the potential risks becomes even more important. To provide a stronger data base on which to compare these materials more fully, the following research agenda is recommended:

- <u>Long-Term Animal-Inhalation Studies</u>. The primary route of human exposure to the seven termiticides is inhalation, and there are only minimal human or animal data on this route. It is possible that there will be variation between ingestion and inhalation, e.g., in pharmacokinetics, dose to critical organ, and toxicologic end points. Differences between routes of administration could modify the relative or absolute risks of these materials.

  Therefore, the Committee recommends long-term inhalation studies of these compounds. Biologic end points to investigate include neurotoxicity, carcinogenicity, effects on blood-forming tissues, and teratogenic and reproductive effects. Studies on the mechanisms of carcinogenicity (particularly the cyclodienes) and neurotoxicity of these compounds should also be undertaken. In addition, the role of metabolism in influencing the toxic effects of the termiticides needs to be examined further. If resources are insufficient for the immediate testing of all seven materials, the Committee suggests that the first studies be done with chlordane, aldrin, and some of the noncyclodiene compounds, such as lindane and chlorpyrifos. Heptachlor and dieldrin would be expected to yield results qualitatively similar to those of chlordane and aldrin. The Committee also suggests investigation of the possibility of undertaking these studies jointly with other government agencies that are responsible for testing environmental chemicals.

- <u>Airborne Monitoring</u>. An important consideration in assessing the risks of the termiticides is knowledge of the airborne concentrations in residences after application. Published quantitative analytic data were available to the Committee only on chlordane; some preliminary data were available on aldrin and dieldrin. The Committee suggests a program to determine the airborne concentrations of the termiticides under conditions similar to those now found in military housing. Monitoring should be conducted over an extended period to delineate the effects of time, temperature, and other variables on airborne concentrations of the termiticides.

• <u>Epidemiology Data</u>. The Committee (NRC, 1979) has previously suggested an epidemiologic study of persons who have been exposed to chlordane in military housing. This group can be followed more readily than the general population, because their health records and whereabouts are more easily traced. The Committee now recommends that, at a minimum, a biologic monitoring program be undertaken as a first step in a comprehensive analysis of the human health effects of chlordane. Concentrations of chlordane and its metabolites in fat, blood, and urine of persons known to have lived in military housing where chlordane was applied should be measured. If data are available, these concentrations before and after exposure should be compared, to learn whether there is appreciable accumulation of chlordane in the body. Observational data on the health status of these persons should also be obtained. In particular, investigators should look for signs of neurotoxicity--such as seizures, movement disorders, tremors, and chorea--and for signs of anemia and diseases of blood-forming tissues. Neurologic symptoms appear to be the most sensitive indicator of exposure in humans. It might be possible to develop a retrospective case-control study of those with suggestive neurologic symptoms and appropriate matched controls (neighbors, unaffected siblings, etc.), including correlation with concentrations of the termiticides in tissues and in indoor air. Any of the other termiticides for which there are sufficient data on exposure in military housing should be investigated in a similar fashion.

Table 1. Physical and Chemical Properties

Aldrin

Chemical name: 1,2,3,4,10,10-Hexachloro-1,4,4a,5,8,8a-hexahydro-1,4,5,8-dimethanonaphthalene
Molecular weight: 365
Molecular formula: $C_{12}H_8Cl_6$
Physical state: Brown to white crystalline solid
Melting point: 104°C
Vapor pressure: 0.000006 mm Hg (at 20°C)
Soluble in most organic solvents
Insoluble in water

Chlordane

Chemical name: 1,2,4,5,6,7,8,8-Octachloro-4,7-methano-3a,4,7,7a-tetrahydroindane
Molecular weight: 410
Molecular formula: $C_{10}H_6Cl_8$
Physical state: Colorless, odorless, viscous liquid
Vapor pressure: 0.00001 mm Hg (at 20°C)
Specific gravity: 1.57-1.67
Soluble in many organic solvents
Solubility in water: 9 µg/L

Chlorpyrifos

Chemical name: O,O-Diethyl O-(3,5,6-trichloro-2-pyridyl) phosphorothioate
Molecular weight: 351
Molecular formula: $C_9H_{11}Cl_3NO_3PS$
Physical state: White crystalline solid
Melting point: 42.5-43°C
Vapor pressure: 1.87×10^{-5} mm Hg (at 25°C)
Soluble in most organic solvents
Solubility in water: 0.0002 g/100 g

Dieldrin

Chemical name: 1,2,3,4,10,10-Hexachloro-6,7-epoxy-1,4,4a,5,6,7,8,8a-octahydro-1,4,5,8-dimethanonaphthalene
Molecular weight: 381
Molecular formula: $C_{12}H_8Cl_6O$
Physical state: Light-tan, flaked, odorless solid
Melting point: 175-176°C
Vapor pressure: 1.8×10^{-7} mm Hg (at 20°C)
Moderately soluble in aromatic hydrocarbons, halogenated solvents, esters, and ketones
Insoluble in water, methanol, and aliphatic hydrocarbons

Heptachlor

Chemical name:	1,4,5,6,7,8,8a-Heptachloro-3a,4,7,7a-tetrahydro-4,7-methanoindane
Molecular weight:	373
Molecular formula:	$C_{10}H_5Cl_7$
Physical state:	White to light tan, waxy solid with a mild camphoraceous odor
Melting point:	95-96°C
Boiling point:	175°C (at 2 mm Hg)
Vapor pressure:	0.0003 mm Hg (at 20°C)
Specific gravity:	1.57-1.59

Soluble in xylene and alcohol
Insoluble in water

Lindane

Chemical name:	Hexachlorocyclohexane
Molecular weight:	291
Molecular formula:	$C_6H_6Cl_6$
Physical state:	White crystalline substance with a slight musty odor
Melting point:	112.5°C
Vapor pressure:	0.14 mm Hg (at 40°C)

Soluble in chloroform, alcohol, acetone, ether, and benzene
Insoluble in water

Pentachlorophenol

Molecular weight:	266
Molecular formula:	C_6HCl_5O
Physical state:	White crystal or powder with a phenolic odor and a pungent taste
Melting point:	190°C
Boiling point:	310°C
Vapor pressure:	0.00017 mm Hg (at 20°C)
Specific gravity:	1.978 (at 20°C)

Soluble in alcohol, acetone, ether, pine oil, and benzene
Slightly soluble in water

Table 2. Deaths, By Major Cause, in White Males Employed for Various Periods between January 1, 1946, and December 31, 1979, at Velsicol Chemical Corporation, Marshall, Illinois, Compared with Expected Deaths, Based on U.S. Population Rates[a]

Minimal Period of Employment	All Causes		Malignant Neoplasms (Cancers)		Cardiovascular (Heart Disease)		Cerebrovascular (Stroke)		Trauma (External Causes)		Other, Unknown	
	Vels.	U.S.	Vels.	U.S.	Vels.	U.S.	Vels.	U.S.	Vels.	U.S.	Vels.	U.S.
3 Mo	120[b]	142.6	22	27.9	64	62.7	5	9.2	14	15.5	15[c]	27.3
1 Yr	104[c]	123.8	21	24.2	54	55.1	4	8.3	14	12.7	11[b]	23.5
5 Yr	74	81.8	16	16.5	39	37.5	4	5.7	8	6.7	7[c]	15.3
10 Yr	56	65.3	14	13.6	30	30.3	2	4.6	5	4.6	5[c]	12.2
15 Yr	39	44.9	11	9.8	19	20.7	1	2.9	4	3.0	4	8.5
20 Yr	25	28.8	7	6.6	11	13.2	0	1.8	4	1.8	3	5.5

[a]Reprinted with permission from Shindell and Associates, 1980.
[b]Statistically significant, $p < 0.01$, expected is greater than observed.
[c]Statistically significant, $p < 0.05$, expected is greater than observed.

Table 3. Standard Mortality Ratios[a] in White Males Employed for Various Periods between January 1, 1946, and December 31, 1979, at Velsicol Chemical Corporation, Marshall, Illinois[b]

Minimal Periods of Employment	Deaths from All Causes	Deaths from Cancer	Deaths from Heart Disease	Deaths from All Other Causes
3 mo	0.84[c]	0.79	1.02	0.65
1 yr	0.84[d]	0.87	0.98	0.65
5 yr	0.90	0.97	1.04	0.68
10 yr	0.86	1.03	0.99	0.56
15 yr	0.87	1.12	0.92	0.62
20 yr	0.87	1.06	0.83	0.78

[a]Standard mortality ratio = ratio of observed to expected, i.e., (no. observed Marshall deaths)/(no. calculated expected deaths).
[b]Reprinted with permission from Shindell and Associates, 1980.
[c]Statistically significant, $p < 0.01$.
[d]Statistically significant, $p < 0.05$.

Table 4. Summary Data on Seven Pesticides

BIOLOGIC END POINTS

Pesticide	Oral LD_{50} for Rats, mg/kg	Observed Critical Effects in Animals	ED_{10} for Mice, ppm	Observed Critical Target Site for Human Exposure
Aldrin[a]	46-63	CNS symptoms, hepatoma	3.1	Central nervous system
Dieldrin[a]	38-52	CNS symptoms, hepatoma	3.6	Central nervous system
Chlordane[a]	335-430	CNS symptoms, hepatoma	16	Central nervous system
Heptachlor[a]	100-160	CNS symptoms, hepatoma	5	Central nervous system
Lindane[b]	125-230	CNS symptoms, liver toxicity	103	Central nervous system
Pentachlorophenol[b]	146-174	CNS symptoms, embryotoxicity	--	Central nervous system
Chlorpyrifos[c]	138-245	Cholinesterase inhibition	--	Plasma cholinesterase

ENVIRONMENTAL END POINTS

Pesticide	Volatility, mm Hg	Reported Airborne Concentrations, $\mu g/m^3$	Amount of Active Ingredient,[d] lb/gal	Persistence After Application, yr
Aldrin[a]	0.000006 (at 20°C)	e	0.04	20
Dieldrin[a]	0.00000018 (at 20°C)	e	0.024	20
Chlordane[a]	0.00001 (at 20°C)	<1-264	0.08	20
Heptachlor[a]	0.0003 (at 20°C)	e	0.04	20
Lindane[b]	0.14 (at 40°C)	e	0.064	10
Pentachlorophenol[b]	0.00017 (at 20°C)	e	0.4	3
Chlorpyrifos[c]	0.0000187 (at 25°C)	e	0.08	4-10

[a]Chlorinated cyclodiene.
[b]Chlorinated hydrocarbon.
[c]Organophosphate.
[d]Estimated amount applied for subslab injection in preexisting structures.
[e]None reported in the literature.

Table 5. Proportions of Male B6C3F1 Mice with Hepatocellular Carcinoma after Pesticide Feeding[a]

Pesticide	Dietary Concentration, ppm	No. Animals with Tumors	No. Animals Examined
Aldrin	0	17	92
	4	16	49
	8	25	46
Chlordane	0	17	92
	30	16	48
	56	43	49
Dieldrin	0	17	92
	2.5	12	50
	5.0	16	45
Heptachlor	0	17	92
	6	11	46
	14	34	47
Lindane	0	5	49
	80	19	49
	160	9	46

[a] Data from NCI, 1977a,b,c; 1978a.

Table 6. Estimates of ED_{10} for Hepatocellular Carcinoma in Male B6C3F1 Mice After Pesticide Feeding

Pesticide	ED_{10}, ppm	95 Percent Confidence Limits
Aldrin	3.1	1.5-6.5
Chlordane	16	14-20
Dieldrin	3.6	2.0-6.6
Heptachlor	5.0	4.1-6.2
Lindane	103	47-225

Table 7. Upper Bounds on Carcinogenic Risk of Pesticides Consumed in Drinking Water[a]

Pesticide	Upper 95 Percent Confidence Estimate of Lifetime Cancer Risk per Microgram per Liter
Dieldrin	260×10^{-6}
Heptachlor	42×10^{-6}
Chlordane	18×10^{-6}
Lindane	9.3×10^{-6}

[a] Data from NRC, 1977b.

REFERENCES

Abbott, D.C., Collins, G.B., and Goulding, R. 1972. Organochlorine pesticide residues in human fat in the United Kingdom 1969-71. Br. Med. J. 2:553-556.

Ahlborg, U.G., Lindgren, J.-E., and Mercier, M. 1974. Metabolism of pentachlorophenol. Arch. Toxicol. 32:271-281.

Ahlborg, U.G., and Thunberg, T. 1978. Effects of 2,3,7,8-tetrachloro-dibenzo-p-dioxin on the in vivo and in vitro dechlorination of pentachlorophenol. Arch. Toxicol. 40:55-61.

Ahmed, F.E., Hart, R.W., and Lewis, N.J. 1977. Pesticide induced DNA damage and its repair in cultured human cells. Mutat. Res. 42:161-174.

Aldrich, F.D., and Holmes, J.H. 1969. Acute chlordane intoxication in a child. Case report with toxicological data. Arch Environ. Health 19:129-132.

Altman, P.L., and Dittmer, D.S., eds. 1974. Biology Data Book. Vol. III, 2nd edition. Bethesda, Maryland: Federation of American Societies for Experimental Biology. p. 1433-2123.

Alvarez, W.C., and Hyman, S. 1953. Absence of toxic manifestations in workers exposed to chlordane. A.M.A. Arch. Ind. Hyg. Occup. Med. 8:480-483.

Ambrose, A.M., Christensen, H.C., and Robbins, D.J. 1953. Pharmacological observations on chlordane. Fed. Proc., Fed. Am. Soc. Exp. Biol. 12:298, abstr. no. 982.

American Conference of Governmental Industrial Hygienists. 1980. Documentation of the Threshold Limit Values, Fourth Edition. Cincinnati, OH: American Conference of Governmental Industrial Hygienists, Inc. 486 p.

American Conference of Governmental Industrial Hygienists. 1981. Threshold Limit Values for Chemical Substances and Physical Agents in the Workroom Environment with Intended Changes for 1981. Cincinnati, OH: ACGIH. 94 p.

Armstrong, R.W., Eichner, E.R., Klein, D.E., Barthel, W.F., Bennett, J.V., Jonsson, V., Bruce, H., and Loveless, L.E. 1969. Pentachlorophenol poisoning in a nursery for newborn infants. II. Epidemiologic and toxicologic studies. J. Pediatr. 75:317-325.

Arnold, D.W., Kennedy, G.L., Jr., Keplinger, M.L., Calandra, J.C., and Calo, C.J. 1977. Dominant lethal studies with technical chlordane, HCS-3260, and heptachlor: heptachlor epoxide. J. Toxicol. Environ. Health 2:547-555.

Avar, P., and Czegledi-Janko, G. 1970. Occupational exposure to aldrin: clinical and laboratory findings. Br. J. Ind. Med. 27:279-282.

Baader, E.W., and Bauer, H.J. 1951. Industrial intoxication due to pentachlorophenol. Ind. Med. Surg. 20:286-290.

Bakke, J.E., Feil, V.J., and Price, C.E. 1976. Rat urinary metabolites from $\underline{O},\underline{O}$-diethyl-$\underline{O}$-(3,5,6-trichloro-2-pyridyl) phosphorothioate. J. Environ. Sci. Health B11:255-230.

Baldwin, M.K., Robinson, J., and Parke, D.V. 1972. A comparison of tne metabolism of HEOD (dieldrin) in the CF1 mouse with tnat in the CFE rat. Food Cosmet. Toxciol. 10:333-351.

Bann, J.M., DeCino, T.J., Earle, N.W., and Sun, Y-P. 1956. The fate of aldrin and dieldrin in the animal body. J. Agric. Food Chem. 4:937-941.

Barnes, R. 1967. Poisoning by the insecticide chlordane. Med. J. Aust. 1:972-973.

Barnett, J.R., and Dorough, H.W. 1974. Metabolism of chlordane in rats. J. Agric. Food Chem. 22:612-619.

Baumann, K., Angerer, J., Heinrich, R., and Lehnert, G. 1980. Occupational exposure to hexachlorocyclohexane. I. Body burden of HCH-isomers. Int. Arch. Occup. Environ. Health 47:119-127.

Baumann, K., Behling, K., Brassow, H.-L., and Stapel, K. 1981. Occupational exposure to hexachlorocyclohexane. III. Neurophysiological findings and neuromuscular function in chronically exposed workers. Int. Arch. Occup. Environ. Health 48:165-172.

Begley, J., Reichert, E.L., Rashad, M.N., Klemmer, H.W., and Siemsen, A.W. 1977. Association between renal function tests and pentachlorophenol exposure. Clin. Toxicol. 11:97-106.

Bergner, H., Constantinidis, P., and Martin, J.H. 1965. Industrial pentachlorophenol poisoning in Winnipeg. Can. Med. Assoc. J. 92:448-451.

Bidwell, K., Weber, E., Nienhold, I., Connor, T., and Legator, M.S. 1975. Comprehensive evaluation for mutagenic activity of dieldrn. Mutat. Res. 31:314, abstr. no. 15.

Brassow, H.-L., Baumann, K., and Lehnert, G. 1981. Occupational exposure to hexachlorocyclohexane. II. Health conditions of chronically exposed workers. Int. Arch. Occup. Environ. Health 48:81-87.

Braun, W.H., and Sauerhoff, M.W. 1976. The pharmacokinetic profile of pentachlorophenol in monkeys. Toxicol. Appl. Pharmacol. 38:525-533.

Braun, W.H., Young, J.D., Blau, G.E., and Gehring, P.J. 1977. The pharmacokinetics and metabolism of pentachlorophenol in rats. Toxicol. Appl. Pharmacol. 41:395-406.

Brooks, G.T. 1974a. Chlorinated Insecticides. Vol. I. Technology and Application. Cleveland, OH: CRC Press. 249 p.

Brooks, G.T. 1974b. Chlorinated insecticides. Vol. II. Biological and Environmental Aspects. Cleveland, OH: CRC Press. 197 p.

Bushland, R.C., Wells, R.W., and Radeleff, R.D. 1948. Effect on livestock of sprays and dips containing new chlorinated insecticides. J. Econ. Entomol. 41:642-645.

Cerey, K., Izakovic, V., and Ruttkay-Nedecka, J. 1973. Effect of heptachlor on dominant lethality and bone marrow in rats. Mutat. Res. 21:26, abstr. no. 10.

Chadwick, R.W., Chuang, L.T., and Williams, K. 1975. Dehydrogenation: A previously unreported pathway of lindane metabolism in mammals. Pestic. Biochem. Physiol. 5:575-586.

Chernoff, N., Kavlock, R.J., Kathrein, J.R., Dunn, J.M., and Haseman, J.K. 1975. Prenatal effects of dieldrin and photodieldrin in mice and rats. Toxicol. Appl. Pharmacol. 31:302-308.

Coulston, F., Goldberg, L., and Griffin, T. 1972. Safety evaluation of DOWCO 179 in human volunteers. Albany, NY: Albany Medical College, Institute of Experimental Pathology and Toxicology. [Unpublished]. 42 p.

Crump, K.S., Guess, H.A., and Deal, K.L. 1977. Confidence intervals and test of hypotheses concerning dose response relations inferred from animal carcinogenicity data. Biometrics 33:437-451.

Curley, A., and Garrettson, L.K. 1969. Acute chlordane poisoning. Arch. Environ. Health 18:211-215.

Curley, A., Burse, V.W., Jennings, R.W., Villanueva, E.C., Tomatis, L., and Akazaki, K. 1973. Chlorinated hydrocarbon pesticides and related compounds in adipose tissue from people of Japan. Nature 242:338-340.

Davidow, B., and Frawley, J.P. 1951. Tissue distribution, accumulation and elimination of the isomers of benzene hexachloride. Proc. Soc. Exp. Biol. Med. 76:780-783.

Davidow, B., and Radomski, J.L. 1953. Isolation of an epoxide metabolite from fat tissues of dogs fed heptachlor. J. Pharmacol. Exp. Ther. 107:259-265.

Davis, K.J., and Fitzhugh, O.G. 1962. Tumorigenic potential of aldrin and dieldrin for mice. Toxicol. Appl. Pharmacol. 4:187-189.

Deacon, M.M., Murray, J.S., Pilny, M.K., Rao, K.S., Dittenber, D.A., Hanley, T.R., Jr., and John, J.A. 1980. Embryotoxicity and fetotoxicity of orally administered chlorpyrifos in mice. Toxicol. Appl. Pharmacol. 54:31-40.

Deichmann, W.B. 1972. Toxicology of DDT and related chlorinated hydrocarbon pesticides. J. Occup. Med. 14:285-292.

Dix, K.M., Van Der Pauw, C.L., and McCarthy, W.V. 1977. Toxicity studies with dieldrin: Teratological studies in mice dosed orally with HEOD. Teratology 16:57-62.

Dover, M., Dow, M., Eckerman, D., Esworthy, R., Glaze, M., and Kuch, P. 1981. Comparative benefit analysis of the seven chemicals registered for use against subterranean termites. Washington, DC: U.S. Environmental Protection Agency, Office of Pesticide Programs. [Unpublished]. 185 p.

Durham, W.F. 1969. Body burden of pesticides in man. Ann. N.Y. Acad. Sci. 160:183-195.

Eliason, D.A., Cranmer, M.F., von Windeguth, D.L., Kilpatrick, J.W., Suggs, J.E., and Schoof, H.F. 1969. Dursban premises applications and their effect on the cholinesterase levels of spraymen. Mosquito News 29:591-595.

Engst, R., Macholz, R.M., Kujawa, M., Lewerenz, H.-J., and Plass, R. 1976. The metabolism of lindane and its metabolites gamma-2,3,4,5,6-pentachlorocyclohexene, pentachlorobenzene, and pentachlorphenol in rats and the pathways of lindane metabolism. J. Environ. Sci. Health B11:95-117.

Environmental Protection Agency. 1974. Notice of intentions to suspend and findings as to an imminent hazard. Fed. Register 39:37246-37272.

Environmental Protection Agency. 1975a. Interim Primary Drinking Water Standards. Fed. Register 40:11990-11998.

Environmental Protection Agency. 1975b National Interim Primary Drinking Water Regulations. Fed. Register 40:59566-59588.

Environmental Protection Agency. 1979a. Chlordane: Ambient Water Quality Criteria. Washington, DC: U.S. Environmental Protection Agency. [67 p.]

Environmental Protection Agency. 1979b. Heptachlor: Ambient Water Quality Criteria. Washington, D.C.: U.S. Environmental Protection Agency. [113 p.]

Environmental Protection Agency. 1979c. Aldrin/Dieldrin: Ambient Water Quality Criteria. Washington, DC: U.S. Environmental Protection Agency. [146 p.]

Environmental Protection Agency. 1979d. Hexachlorocyclohexane: Ambient Water Quality Criteria. Washington, D.C.: U.S. Environmental Protection Agency. [105 p.]

Environmental Protection Agency. 1979e. Pentachlorophenol: Ambient Water Quality Criteria. Washington, DC: U.S. Environmental Protection Agency. [90 p.]

Epstein, S.S. 1976. Carcinogenicity of heptachlor and chlordane. Sci. Total Environ. 6:103-154.

Feldmann, R.J., and Maibach, H.I. 1974. Percutaneous penetration of some pesticides and herbicides in man. Toxicol. Appl. Pharmacol. 28:126-132.

Fishbein, W.J., White, J.V., and Isaacs, H.J. 1964. Survey of workers exposed to chlordane. Ind. Med. Surg. 33:726-727.

Fitzhugh, O.G., Nelson, A.A., and Quaife, M.L. 1964. Chronic oral toxicity of aldrin and dieldrin in rats and dogs. Food Cosmet. Toxicol. 2:551-562.

Freal, J.J., and Chadwick, R.W. 1973. Metabolism of hexachlorocyclohexane to chlorophenols and effect of isomer pretreatment on lindane metabolism in rat. J. Agric. Food Chem. 21:424-427.

Frings, H., and O'Tousa, J.E. 1950. Toxicity to mice of chlordane vapor and solutions administered cutaneously. Science 111:658-660.

Gaines, T.B. 1969. Acute toxicity of pesticides. Toxicol. Appl. Pharmacol. 14:515-534.

Gannon, N., and Bigger, J.H. 1958. The conversion of aldrin and heptachlor to their expoxides in soil. J. Econ. Entom. 51:1-2.

Gannon, N., and Decker, G.C. 1958a. The conversion of heptachlor to its epoxide on plants. J. Econ. Entom. 51:3-7.

Gannon, N., and Decker, G.C. 1958b. The conversion of aldrin to dieldrin on plants. J. Econ. Entomol. 51:8-11.

Garrettson, L.K., and Curley, A. 1969. Dieldrin: Studies in a poisoned child. Arch. Environ. Health 19:814-822.

[General Accounting Office], Comptroller General of the United States, Washington, D.C. 1980. Need for a Formal Risk/Benefit Review of the Pesticide Chlordane, (CED-80-116). [Rept. no.] B-199618. Aug. 5. 11 p.

Goldstein, J.A., Friesen, M., Linder, R.E., Hickman, P., Hass, J.R., and Bergman, H. 1977. Effects of pentachlorophenol on hepatic drug-metabolizing enzymes and porphyria related to contamination with chlorinated dibenzo-p-dioxins and dibenzofurans. Biochem. Pharmacol. 26:1549-1557.

Gosselin, R.E., Hodge, H.C., Smith, R.P., and Gleason, M.N. 1976. Clinical Toxicology of Commercial Products. Acute Poisoning. 4th ed. Baltimore: Williams and Wilkins.

Gray, H.E. 1965. DURSBAN... A new organo-phosphorus insecticide. Down to Earth 21:2, 26-27.

Gutenmann, W.H., St. John, L.E., Jr., and Lisk, D.J. 1968. Metabolic studies with $\underline{O},\underline{O}$-diethyl $\underline{O}$-(3,5,6-trichloro-2-pyridyl) phosphorothioate (Dursban) insecticide in a lactating cow. J. Agric. Food Chem. 16:45-47.

Harrington, J.M., Baker, E.L., Jr., Folland, D.S., Saucier, J.W., and Sandifer, S.H. 1978. Chlordane contamination of a municipal water system. Environ. Res. 15:155-159.

Hayes, W.J., Jr. 1957. Dieldrin poisoning in man. Publ. Health Rep. 72:1087-1091.

Hayes, W.J., Jr. 1959. The toxicity of dieldrin to man. Bull. World Health Org. 20:891-912.

Hayes, W.J., Jr., and Curley, A. 1968. Storage and excretion of dieldrin and related compounds. Effect of occupational exposure. Arch. Environ. Health 15:155-162.

Heath, D.F., and Vandekar, M. 1964. Toxicity and metabolism of dieldrin in rats. Br. J. Ind. Med. 21:269-279.

Hinkle, D.K. 1973. Fetotoxicity effects of pentachlorophenol in the golden Syrian hamster. Toxicol. Appl. Pharmacol. 25:455, abstr. no. 42.

Hoogendam, I., Versteeg, J.P.J., and de Vlieger, M. 1962. Electroencephalograms in insecticide toxicity. Arch. Environ. Health 4:86-94.

Hoogendam, I., Versteeg, J.P.J., and de Vlieger, M. 1965. Nine years' toxicity control in insecticide plants. Arch. Environ. Health 10:441-448.

Hunter, C.G., and Robinson, J. 1967. Pharmacodynamics of dieldrin (HEOD). I. Ingestion by human subjects for 18 months. Arch. Environ. Health 15:614-626.

Hunter, C.G., Robinson, J., and Roberts, M. 1969. Pharmacodynamics of dieldrin (HEOD). Ingestion by human subjects for 18 to 24 months, and postexposure for eight months. Arch. Environ. Health 18:12-21.

Infante, P.F., Epstein, S.S., and Newton, W.A., Jr. 1978. Blood dyscrasias and childhood tumors and exposure to chlordane and heptachlor. Scand. J. Work Environ. Health 4:137-150.

Ingle, L. 1952. Chronic oral toxicity of chlordan to rats. Arch. Ind. Hyg. Occup. Med. 6:357-367.

Ingle, L. 1953. The toxicity of chlordane vapors. Science 118:213-214.

Innes, J.R.M., Ulland, B.M., Valerio, M.G., Petrucelli, L., Fishbein, L., Hart, E.R., Pallotta, A.J., Bates, R.R., Falk, H.L., Gart, J.J., Klein, M., Mitchell, I., and Peters, J. 1969. Bioassay of pesticides and industrial chemicals for tumorigenicity in mice: A preliminary note. J. Nat. Cancer Inst. 42:1101-1114.

International Agency for Research on Cancer. 1974. IARC Monographs on the Evaluation of the Carcinogenic Risk of Chemicals to Man: Some Organochlorine Pesticides. Vol. 5. Lyon: International Agency for Research on Cancer. 241 p.

International Agency for Research on Cancer. 1979. IARC Monographs on the Evaluation of the Carcinogenic Risk of Chemicals to Humans: Some Halogenated Hydrocarbons. Vol. 20. Lyon: International Agency for Research on Cancer. 609 p.

Jager, K.W. 1970. Aldrin, Dieldrin, Endrin and Telodrin: An Epidemiological and Toxicological Study of Long-Term Occupational Exposure. New York: Elsevier Publishing Co. 234 p.

Johnson, R.L., Gehring, P.J., Kociba, R.J., and Schwetz, B.A. 1973 Chlorinated dibenzodioxins and pentachlorophenol. Environ. Health Perspect. 5:171-175.

Kacew, S., Sutherland, D.J.B., and Singhal, R.L. 1973. Biochemical changes following chronic administration of heptachlor, heptachlor epoxide and endrin to male rats. Environ. Physiol. Biochem. 3:221-229.

Kazen, C., Bloomer, A., Welch, R., Oudbier, A., and Price, H. 1974. Persistence of pesticides on the hands of some occupationally exposed people. Arch. Environ. Health 29:315-318.

Khera, K.S., Whalen, C., Trivett, G., and Angers, G. 1979. Assessment of the teratogenic potential of biphenyl, ethoxyquin, piperonyl butoxide, diuron, thiabendazole, phosalone, and lindane in rats. Toxicol. Appl. Pharmacol. 48:A33, abstr. No. 66.

Kilian, D.J., Edwards, H.N., Benge, M., and Tabatabai, Z. 1970. Results of human skin exposure to DOWCO 179. Texas: The Dow Chemical Co. Industrial Medicine and Toxicology Department. [12 p.]

Kimbrough, R.D., and Linder, R.E. 1975. The effect of technical and 99 percent pure pentachlorophenol on the rat liver. Light microscopy and ultrastructure. Toxicol. Appl. Pharmacol. 33:131-132, abstr. No. 23.

Klimmer, O.R. 1955. Experimental studies on the toxicology of insecticidal chlorinated hydrocarbon. Arch. Exp. Pathol. Pharmakol. 227:183-195. [CA 50:3640d, 1956].

Knudsen, I., Verschuuren, H.G., Den Tonkelaar, E.M., Kroes, R., and Helleman, P.F.W. 1974. Short-term toxicity of pentachlorophenol in rats. Toxicology 2:141-152.

Larsen, R.V., Born, G.S., Kessler, W.V., Shaw, S.M., and Van Sickle, D.C. 1975. Placental transfer and teratology of pentachlorophenol in rats. Environ. Lett. 10:121-128.

Lehman, A.J. 1952. Chemicals in foods: A report to the Association of Food and Drug Officials on current developments. Part II. Pesticides. Section III. Subacute and chronic toxicity. Assoc. Food Drug Off. U.S., Q. Bull. 16:47-53.

Lillie, T.H. 1981. Chlordane in Air Force family housing: A study of houses treated after construction. OEHL Report no. 81-45. USAF Occupational and Environmental Health Laboratory, Brooks Air Force Base, TX.

Livingston, J.M., and Jones, C.R. 1981. Living area contamination of chlordane used for termite treatment. Bull. Environ. Contam. Toxicol. 27:406-411.

Lu, P.Y., Metcalf, R.L., Hirwe, A.S., and Williams, J.W. 1975. Evaluation of environmental distribution and fate of hexachlorocyclopentadiene, chlordene, heptachlor, and heptachlor epoxide in a laboratory model ecosystem. J. Agric. Food Chem. 23:967-973.

Ludwig, P.D., Kilian, D.J., Dishburger, H.J., and Edwards, H.N. 1970. Results of human exposure to thermal aerosols containing DURSBAN insecticide. Mosquito News 30:346-354.

Majumdar, S.K., Maharam, L.G., and Viglianti, G.A. 1977. Mutagenicity of dieldrin in the Salmonella-microsome test. J. Hered. 68:184-185.

Marshall, T.C., Dorough, H.W., and Swim, H.E. 1976. Screening of pesticides for mutagenic potential using *Salmonella typhimurium* mutants. J. Agric. Food Chem. 24:560-563.

Matsumura, F., Patil, K.C., Boush, G.M. 1970. Formation of "photodieldrin" by microorganisms. Science 170:1206-1207.

McCollister, S.B., Kociba, R.J., Humiston, C.G., McCollister, D.D., and Gehring, P.J. 1974. Studies of the acute and long-term oral toxicity of chlorpyrifos ($\underline{O},\underline{O}$-diethyl-$\underline{O}$-(3,5,6-trichloro-2-pyridyl) phosphorothioate). Food Cosmet. Toxicol. 12:45-61.

McConnell, E.E., Moore, J.A., Gupta, B.N., Rakes, A.H., Luster, M.I., Goldstein, J.A., Haseman, J.K., and Parker, C.E. 1980. The chronic toxicity of technical and analytical pentachlorophenol in cattle. I. Clinicopathology. Toxicol. Appl. Pharmacol. 52:468-490.

McKellar, R.L., Dishburger, H.J., Rice, J.R., Craig, L.F., and Pennington, J. 1976. Residues of chlorpyrifos, its oxygen analogue, and 3,5,6-trichloro-2-pyridinol in milk and cream from cows fed chlorpyrifos. J. Agric. Food Chem. 24:283-286.

Melis, R. 1955. Tolerability of small doses of lindane by warm-blood animals. I. Effects of lindane administered two to ten parts per million to white rats. Nuovi Ann. Ig. Microbiol. 6:90-104.

Menon, J.A. 1958. Tropical hazards associated with the use of pentachlorophenol. Br. Med. J. 1:1156-1158.

Mestitzova, M. 1967. On reproduction studies and the occurrence of cataracts in rats after long-term feeding of the insecticide heptachlor. Experientia 23:42-43.

Miyazaki, S., and Hodgson, G.C. 1972. Chronic toxicity of Dursban and its metabolite, 3,5,6-trichloro-2-pyridinol in chickens. Toxicol. Appl. Pharmacol. 23:391-398.

Morgan, D.P., Roberts, R.J., Walter, A.W., and Stockdale, E.M. 1980. Anemia associated with exposure to lindane. Arch. Environ. Health 35:307-310.

Nagasaki, H., Tomii, S., Mega, T., Marugami, M., and Ito, N. 1972a. Hepatocarcinogenic effect of α-, β-, γ-, and δ-isomers of benzene hexachloride in mice. Gann. 63:393.

Nagasaki, H., Tomii, S., Tsumashika, T., Marukami, M., Arai, M., and Ito, N. 1972b. On the experimental tumorigenesis of the liver of mice and rats by the induction of BHC isomers, α-, β-, γ-, and δ. Nippon Gangakkai, Kijii 31:33. (Health Aspects Pestic. 6:388-389, abstr. no. 73-1720).

Naishtein, S.Ya., and Leibovich, D.L. 1971. Effect of small doses of DDT and lindane and their mixture on sexual function and embryogenesis in rats. Hyg. Sanit. 36:190-195.

Nantel, A.J., Ayotte, L., Beneditti, J.-L., Savoie, J.-Y., Tessier, L., Schwarz, T., and Weber, J.-P. 1977. A group of adults acutely poisoned by food contaminated with lindane. Acta Pharmacol. Toxicol. 41(Suppl. 2):250.

National Cancer Institute, Division of Cancer Cause and Prevention. 1977a. Bioassay for Chlordane for Possible Carcinogenicity. NCI-CG-TR-8. Bethesda, MD: National Institutes of Health. 117 p. [DHEW Publ. No. (NIH) 77-808].

National Cancer Institute, Division of Cancer Cause and Prevention. 1977b. Bioassay of Heptachlor for Possible Carcinogenicity. NCI-CG-TR-9. Bethesda, MD: National Institutes of Health. 111 p. [DHEW Publ. No. (NIH) 77-809].

National Cancer Institute, Division of Cancer Cause and Prevention. 1977c. Bioassay of Lindane for Possible Carcinogenicity. NCI-CG-TR-14. Bethesda, MD: National Institutes of Health. 99 p. [DHEW Publ. No. (NIH) 77-814].

National Cancer Insitute, Division of Cancer Cause and Prevention. 1978a. Bioassays of Aldrin and Dieldrin for Possible Carcinogenicity. NCI-CG-TR-21. Bethesda, MD: National Institutes of Health. 184 p. [DHEW Publ. No. (NIH) 78-821].

National Cancer Institute, Divison of Cancer Cause and Prevention. 1978b. Bioassay of Dieldrin for Possible Carcinogenicity. NCI-CG-TR-22. Bethesda, MD: National Institutes of Health. 50 p. [DHEW Publ. No. (NIH) 78-822].

National Institute for Occupational Safety and Health, Center for Disease Control, Public Health Service. 1978. Special Occupational Hazard Review for Aldrin/Dieldrin. Rockville, Maryland: U.S. Department of Health, Education, and Welfare. 166 p. [DHEW (NIOSH) Publ. No. 78-201].

National Research Council, Assembly of Life Sciences, Advisory Center on Toxicology, Pesticide Information Review and Evaluation Committee. 1977a. An Evaluation of the Carcinogenicity of Chlordane and Heptachlor. Washington, DC: National Academy of Sciences. [118 p.]

National Research Council, Assembly of Life Sciences, Advisory Center on Toxicology, Safe Drinking Water Committee. 1977b. Drinking Water and Health. Washington, DC: National Academy of Sciences. 939 p.

National Research Council, Committee on Toxicology. 1978. Chlorpyrifos - Risk Assessment and Inhalation Exposure Limits. Washington, DC: National Academy of Sciences.

National Research Council, Committee on Toxicology. 1979. Chlordane in Military Housing. Washington, DC: National Academy of Sciences.

Nisbet, I.C.T. 1976. Human exposure to chlordane, heptachlor, and their metabolites. A review prepared for the Cancer Assessment Group. Washington, DC: U.S. Environmental Protection Agency.

Nye, D.E., and Dorough, H.W. 1976. Fate of insecticides administered endotracheally to rats. Bull. Environ. Contam. Toxicol. 15:291-296.

Occupational Safety and Health Administration. 1981. Air Contaminants. 29 CFR 1910:1000.

Oshiba, K. 1972. Experimental studies on the fate of β- and γ-BHC in vivo following daily administration. Osaka Shiritsu Daigaku Igaku Zasshi (J. Osaka City Med. Cent.) 21:1-9. [English summary]. [Pestic. Abstr. 6:148, No. 73-0682, 1973.]

Ottolenghi, A.D., Haseman, J.K., and Suggs, F. 1974. Teratogenic effects of aldrin, dieldrin, and endrin in hamsters and mice. Teratology 9:11-16.

Palmer, A.K., Bottomley, A.M., Worden, A.N., Frohberg, H., and Bauer, A. 1978a. Effect of lindane on pregnancy in the rabbit and rat. Toxicology 9:239-247.

Palmer, A.K., Cozens, D.D., Spicer, E.J.F., and Worden, A.N. 1978b. Effects of lindane upon reproductive function in a 3-generation study in rats. Toxicology 10:45-54.

Patel, T.B., and Rao, V.N. 1958. "Dieldrin" poisoning in man. A report of 20 cases observed in Bombay State. Br. Med. J. 1:919-921.

Pennington, J.Y., and Edwards, H.N. 1971. Comparison of cholinesterase depression in humans and rabbits following exposure to chlorpyrifos. Lake Jackson, Texas: Dow Chemical, Agricultural Department. [Unpublished]. 22 p.

Polen, P.B., Hester, M., and Benziger, J. 1971. Characterization of oxychlordane, animal metabolite of chlordane. Bull. Environ. Contam. Toxicol. 5:521-528.

Poole, D.C., Simmon, V.F., and Newell, G.W. 1977. In vitro mutagenic activity of fourteen pesticides. Toxicol. Appl. Pharmacol. 41:196, abstr. no. 155.

Princi, F., and Spurbeck, G.H. 1951. A study of workers exposed to the insecticides chlordan, aldrin, dieldrin. A.M.A. Arch. Ind. Hyg. Occup. Med. 3:64-72.

Radomski, J.L., and Davidow, B. 1953. The metabolite of heptachlor, its estimation, storage, and toxicity. J. Pharmacol. Exp. Ther. 107:266-272.

Rao, K.R., ed. 1978. Pentachlorophenol: Chemistry, Pharmacology, and Environmental Toxicology. New York: Plenum Press. 402 p.

Reuber, M.D. 1979. Carcinogenicity of lindane. Environ. Res. 19:460-481.

Schwetz, B.A., Gehring, P.J., and Kociba, R.J. 1973. Toxicological properties of pentachlorophenol relative to its content of chlorinated dibenzo-p-dioxins. Pharmacologist 15:395.

Schwetz, B.A., Keeler, P.A., and Gehring, P.J. 1974. The effect of purified and commercial grade pentachlorophenol on rat embryonal and fetal development. Toxicol. Appl. Pharmacol. 28:151-161.

Schwetz, B.A., Quast, J.F., Keeler, P.A., Humiston, C.G., and Kociba, R.J. 1978. Results of two-year toxicity and reproduction studies on pentachlorophenol in rats. IN: Rao, K. ed. Pentachlorophenol: Chemistry, Pharmacology, and Environmental Toxicology. New York: Plenum Press. p. 301-309.

Shindell and Associates. 1980. Report of Epidemiologic Study of the Employees of Velsicol Chemical Corporation Plant, Marshall, Illinois, January 1946-December 1979. Milwaukee, WI: Shindell and Associates. [Unpublished]. 29 p.

Shirasu, Y., Moriya, M., Kato, K., Furuhashi, A., and Kada, T. 1976. Mutagenicity screening of pesticides in the microbial system. Mutation Res. 40:19-30.

Shirasu, Y., Moriya, M., Kato, K., Lienard, F., Tezuka, H., Teramoto, S., and Kada, T. 1977. Mutagenicity screening on pesticides and modification products: A basis of carcinogenicity evaluation. IN: H.H. Hiatt, J.D. Watson, and J.A. Winsten, eds. Origins of Human Cancer. Book A. Incidence of Cancer in Humans. Cold Spring Harbor Conferences on Cell Proliferation. Vol. 4. New York: Cold Spring Harbor Laboratory. p. 267-285.

Simmon, V.F., Kauhanen, K., and Tardiff, R.G. 1977. Mutagenic activity of chemicals identified in drinking water. IN: D. Scott, B.A. Bridges, and F.H. Sobels, eds. Progress in Genetics Toxicology: Developments in Toxicology and Environmental Sciences. Vol. 2. Proceedings of the Second International Conference on Environmental Mutagens, Edinburgh, July 11-15. New York: Elsevier/North-Holland. p. 249-258.

Smith, G.N., Watson, B.S., and Fischer, F.S. 1967. Investigations on Dursban insecticide. Metabolism of [^{36}Cl] O,O-diethyl-O-3,5,6-trichloro-2-pyridyl phosphorothioate in rats. J. Agric. Food Chem. 15:132-138.

Sperling, F., and Ewinike, H. 1969. Changes in LD_{50} of parathion and heptachlor after turpentine pretreatment. Toxicol. Appl. Pharmacol. 14:622, abstr. no. 24.

Street, J.C., and Blau, S.E. 1972. Oxychlordane: Accumulation in rat adipose tissue on feeding chlordane isomers or technical chlordane. J. Agric. Food Chem. 20:395-397.

Tashiro, S., and Matsumura, F. 1977. Metabolic routes of cis- and trans-chlordane in rats. J. Agric. Food Chem. 25:872-880.

Taylor, J.R., Calabrese, V.P., and Blanke, R. 1979. Organochlorine and other insecticides, chapter 16. IN: Vinken, P.J., and G.W. Bruyn, eds. Handbook of Clinical Neurology, Vol. 36. Intoxications of the Nervous System, Part I. New York: Elsevier/North-Holland. p. 391-455.

Thorpe, E., and Walker, A.I.T. 1973. The toxicology of dieldrin (HEOD). II. Comparative long-term oral toxicity studies in mice with dieldrin, DDT, phenobarbitone, β-BHC and γ-BHC. Food Cosmet. Toxicol. 11:433-442.

Tomczak, S., Baumann, K., and Lehnert, G. 1981. Occupational exposure to hexachlorocyclohexane. IV. Sex hormone alterations in HCH-exposed workers. Int. Arch. Occup. Environ. Health 48:283-287.

Torkelson, T.R. 1965. Chronic inhalation of atmospheres containing O,O-diethyl-O-3,5,6-trichloro-2-pyridyl ester of phosphoric acid (DURSBAN). Midland, MI: Dow Chemical Co., Biochemical Research Laboratory. [3 p.]

Truhaut, R., Gak, J.-C., and Graillot, C. 1974. Research on the modes and mechanisms of toxic action of organochlorinated insecticides. I. Comparative study of the effects of acute toxicity on the hamster and the rat. Trans. of J. Eur. Toxicol. 7:159-166.

Vogel, E. and Chandler, J.L.R. 1974. Mutagenicity testing of cyclamate and some pesticides in Drosophila melanogaster. Experientia 30:621-623.

Walker, A.I.T., Thorpe, E., and Stevenson, D.E. 1973. The toxicology of dieldrin (HEOD). I. Long-term oral toxicity studies in mice. Food Cosmet. Toxicol. 11:415-432.

Wang, H.H., and MacMahon, B. 1979. Mortality of workers employed in the manufacture of chlordane and heptachlor. J. Occup. Med. 21:745-748.

Warner, S.D., Gerbig, C.G., Strebing, R.J., and Molello, J.A. 1980. Results of a two-year toxicity and oncogenic study of chlorpyrifos administered to CD-1 mice in the diet. Indianapolis, IN: Dow Chemical Co. [Unpublished]. 140 p.

Weisse, I., and Herbst, M. 1977. Carcinogenicity study of lindane in the mouse. Toxicology 7:233-238.

Welch, R.M., Levin, W., Kuntzman, R., Jacobson, M., and Conney, A.H. 1971. Effect of halogenated hydrocarbon insecticides on the metabolism and uterotropic action of estrogens in rats and mice. Toxicol. Appl. Pharmacol. 19:234-246.

World Health Organization, Food and Agricultural Organization of the United Nations. 1967. Evaluation of Some Pesticide Residues in Food. WHO/Food Add./67.32. Joint Meeting of the FAO Working Party and the WHO Expert Committee on Pesticide Residues, Nov 14-21, 1966. Geneva: World Health Organization. 237 p.

World Health Organization, Food and Agricultural Organization of the United Nations. 1968. 1967 Evaluation of Some Pesticide Residues in Food. WHO/Food Add./68.30. Joint Meeting of the FAO Working Party of Experts and the WHO Expert Committee on Pesticide Residues, Dec. 4-11, 1967. Geneva: World Health Organization. 242 p.

World Health Organization. 1973. 1972 Evaluations of Some Pesticide Residues in Food. WHO Pesticide Residues Series No. 2. Geneva: World Health Organization. p. 161-162.

Wright, C.G., and Leidy, R.B. 1982. Chlordane and heptachlor in the ambient air of houses treated for termites. Bull. Environ. Contam. Toxicol. 28:617-623.

BIOGRAPHIC SKETCHES OF COMMITTEE MEMBERS

Donald J. Ecobichon is a Professor of Pharmacology at McGill
University, Montreal, Quebec. He received a B.Sc.Phm. in 1960, an
M.A. in 1962, and a Ph.D. in pharmacology in 1964 from the
University of Toronto. Dr. Ecobichon's research has focused on the
toxicology of chlorinated hydrocarbons and organophosphorus
insecticides and drug hydrolysis by tissue esterases of various
mammalian species. Dr. Ecobichon's professional affiliations
include the Society of Toxicology, the Pharmacology Society of
Canada, and the New York Academy of Sciences.

David W. Gaylor is Director of the Division of Biometry of the National
Center for Toxicological Research, Jefferson, Arkansas. He received
a B.S. in 1951 and an M.S. in 1953 from Iowa State University and a
Ph.D. in statistics in 1960 from North Carolina State University.
He holds a concurrent position as Adjunct Professor at the
University of Arkansas Medical School. Dr. Gaylor is a Fellow of
the American Statistical Association and a member of the Biometry
Society. He is an expert in statistical design and analysis of
experiments and in biostatistics.

Peter Greenwald is Director of Resources, Centers, and Community
Activities of the National Cancer Institute, Bethesda, Maryland. He
received an M.D. in 1961 from the State University of New York at
Syracuse and an M.P.H. in 1967 and a Dr.P.H. in 1974 from Harvard
University. Dr. Greenwald is board-certified in medicine and
preventive medicine. He has served as Director of the Cancer
Control Bureau of the New York State Department of Health. Dr.
Greenwald is an expert in epidemiology.

Ian T. Higgins is Professor of Epidemiology and Professor of
Environmental and Industrial Health at the University of Michigan
School of Public Health, Ann Arbor. He received a B.S. in 1946 and
M.S. in 1951 from the University of London. Dr. Higgins has served
on a number of committees of the National Research Council and is a
member of various professional societies, including the American
Epidemiology Society and the British Thoracic Society. He is an
expert in the epidemiology of chronic diseases, particularly
pneumoconiosis and chronic respiratory diseases.

Wendell W. Kilgore is Professor of Environmental Toxicology
at the University of California, Davis. He is a former chairman of
the department and director of the Food Protection and Toxicology
Center. Dr. Kilgore received an A.B. in 1951 and a Ph.D. in
microbiology in 1959 from the University of California. Among his
affiliations are the American Society of Microbiology and the
Society of Toxicology. His research has focused on insect and
microbial biochemistry and the mechanisms of action of pesticides.

Leonard T. Kurland is Professor of Epidemiology at Mayo Graduate
School of Medicine and Chairman of the Department of Epidemiology
and Medical Statistics at Mayo Clinic, Rochester, Minnesota. He
received a B.A. in 1942 and a Dr.P.H. in 1951 from Johns Hopkins
University, an M.D. from the University of Maryland in 1945, and an
M.P.H. from Harvard University in 1948. Dr. Kurland is a past
president of the American Epidemiology Society, and his professional
affiliations include the American Neurological Association, the
American Public Health Asociation, and the American Society of Human
Genetics. He is an expert in medical record systems and the
epidemiology of chronic disease and human genetics as applied to
neurology.

Howard I. Maibach is Professor of Dermatology at the University
of California, San Francisco, School of Medicine. He received an
A.B. in 1950 and an M.D. in 1955 from Tulane University. Dr.
Maibach is a diplomate of the American Board of Dermatology. Among
his affiliations are the Society of Investigative Dermatology, the
American Academy of Dermatology, and the American Federation of
Clinical Research. Dr. Maibach is an expert in the biologic effects
of dermal exposure to chemicals.

H. George Mandel is Professor and Chairman of the Department of
Pharmacology at the George Washington University School of Medicine,
Washington, D.C. He received a B.S. in 1944 and a Ph.D. in organic
chemistry in 1949 from Yale University. He has served in a variety
of concurrent positions for a number of universities and
organizations both in the United States and abroad, including the
National Cancer Institute, the Veterans' Administration, and the
American Cancer Society. Dr. Mandel is a past president of the
American Society of Pharmacology and Experimental Therapeutics and a
member of the American Association of Cancer Research and the
American Society of Biological Chemistry. His research has been
devoted to drug metabolism and the mechanisms of action of
anticancer drugs.

Roger O. McClellan is Director and Toxicologist of the Inhalation
Toxicology Research Institute, Lovelace Biomedical and Environmental
Research Institute, Albuquerque, New Mexico. Dr. McClellan received
a D.V.M. in 1960 from Washington State University, is a diplomate of
the American Board of Veterinary Toxicology, and is certified by the
American Board of Toxicology. Dr. McClellan is an expert in the
metabolism and toxicity of inhaled materials, especially materials
released from the use of different energy technologies. He has
served on numerous government committees dealing with radiation
protection and environmental health, including those of the
Environmental Protection Agency and the National Institutes of
Health. Dr. McClellan's professional affiliations include the
Society of Toxicology, the American College of Veterinary
Toxicology, the Radiation Research Society, the Health Physics
Society, and the Society of Risk Analysis.

Joseph V. Rodricks is a principal in Environ Corporation,
Washington, D.C. He has been Deputy Associate Commissioner for
Health Affairs at the Food and Drug Administration. Dr. Rodricks
received a B.S. in 1960 from MIT and a Ph.D. in chemistry in 1968
from the University of Maryland. Dr. Rodricks is an expert in
environmental health, focusing on measurement of risks in humans
from exposure to toxic substances. Among his professional
affiliations are the Society of Risk Analysis and the American
College of Toxicology. He is also on the Board of Directors of the
Academy of Toxicological Sciences.

Ronald C. Shank is Associate Professor of Toxicology and Director of
the Program on Environmental Toxicology at the University of
California, Irvine. Dr. Shank received an Sc.B. in 1959, an Sc.M.
in 1961, and a Ph.D. in nutrition and biochemistry in 1964 from
MIT. Dr. Shank has served as Director of the MIT Mycotoxin Research
Project in Southeast Asia. Dr. Shank's research has been devoted to
the alkylation of nucleic acids and proteins from exposure to toxic
substances and the toxicology of foodborne substances. He is a
member of the Society of Toxicology.

Edward A. Smuckler is Professor and Chairman of the Department of
Pathology at the University of California, San Francisco, School of
Medicine. He received an A.B. in 1952 from Dartmouth College, an
M.D. in 1956 from Tufts University, and a Ph.D. in pathology in 1963
from the University of Washington. Dr. Smuckler is a member of
various professional societies, including the American Society of
Experimental Pathology, the American Society of Pathology and
Bacteriology, and the Society of Toxicology. He is an expert in
cellular alterations from exposure to toxic substances and has
served on several government advisory committees on
environmental-health matters.

Robert Snyder is Professor and Chairman of the Department of
Pharmacology at Rutgers University School of Pharmacy and Director
of the Toxicology Program at Rutgers University, Piscataway, New
Jersey. He received a B.S. in 1957 from Queens College and a Ph.D.
in biochemistry in 1961 from the State University of New York. Dr.
Snyder's affiliations include the New York Academy of Sciences, the
American Society of Pharmacology and Experimental Therapeutics, and
the Society of Toxicology. His research has been devoted to drug
metabolism and the relationship of metabolism of xenobiotics to
toxic activity.

Peter Spencer is Associate Professor of Neurosciences and Executive
and Scientific Director of the Institute of Neurotoxicology at
Albert Einstein College of Medicine, Bronx, New York. He received a
B.S. and a Ph.D. in medicine from the University of London. He is a
recipient of the Joseph P. Kennedy, Jr., Fellowship in the
Neurosiences. His research has been devoted to the nervous system

effects of exposure to toxic substances and the study of nerve fiber development, degeneration, and regeneration. Dr. Spencer is a member of various professional societies, including the American Society of Cell Biology, the American Association of Neuropathologists, the Society of Neuroscience, and the Society of Toxicology.

Philip G. Watanabe is a Research Toxicologist at Dow Chemical, U.S.A., Midland, Michigan. He received a B.S. in 1969 from the University of California, Irvine, and a Ph.D. in toxicology in 1974 from Utah State University. His affiliations include the Society of Toxicology and the International Society of Biochemical Pharmacology. Dr. Watanabe's research has focused on the pharmacokinetics of toxic substances and the relationship of pharmacokinetics to biologic activity.